Collins

Student Book

EDEXCEL INTERNATIONAL GCSE (9-1) CHEMISTRY

Sam Goodman and Chris Sunley

William Collins' dream of knowledge for all began with the publication of his first book in 1819. A self-educated mill worker, he not only enriched millions of lives, but also founded a flourishing publishing house. Today, staying true to this spirit, Collins books are packed with inspiration, innovation and practical expertise. They place you at the centre of a world of possibility and give you exactly what you need to explore it.

Collins. Freedom to teach.

Published by Collins
An imprint of HarperCollins*Publishers*
The News Building
1 London Bridge Street
London
SE1 9GF

Browse the complete Collins catalogue at
www.collins.co.uk

Authors: Sam Goodman and Chris Sunley
Original material by Sam Goodman and Chris Sunley
New material by Chris Sunley
Commissioning Editor: Joanna Ramsay
Development Editor: Gillian Lindsey
Project manager: Maheswari PonSaravanan
Project editor: Vicki Litherland
Copy editor: Jan Schubert
Proofreader: Sara Hulse
Indexer: Jane Henley
Answer checker: Aidan Gill
Typesetting: Jouve India
Artwork: Jouve India
Cover design: ink-tank
Production: Rachel Weaver
Printed by: Grafica Veneta

MIX
Paper from
responsible sources
FSC® C007454

This book is produced from independently certified FSC paper to ensure responsible forest management.

For more information visit: **www.harpercollins.co.uk/green**

Contents

Getting the best from the book

Welcome to *Edexcel International GCSE Chemistry*.

This textbook has been designed to help you understand all of the requirements needed to succeed in the Edexcel International GCSE Chemistry course. Just as there are four sections in the Edexcel specification, so there are four sections in the textbook: Principles of chemistry, Inorganic chemistry, Physical chemistry and Organic chemistry.

Each section is split into topics. Each topic in the textbook covers the essential knowledge and skills you need. The textbook also has some very useful features which have been designed to really help you understand all the aspects of chemistry which you will need to know for this specification.

SAFETY IN THE SCIENCE LESSON

This book is a textbook, not a laboratory or practical manual. As such, you should not interpret any information in this book that related to practical work as including comprehensive safety instructions. Your teachers will provide full guidance for practical work and cover rules that are specific to your school.

A brief introduction to the section to give context to the science covered in the section.

Starting points will help you to revise previous learning and see what you already know about the ideas to be covered in the section.

The section contents shows the separate topics to be studied matching the specification order.

Learning objectives cover what you need to learn in this topic.

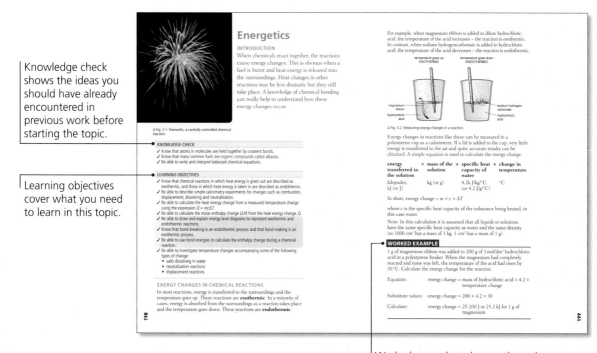

Worked examples take you through how to apply formulae that you need to know how to use.

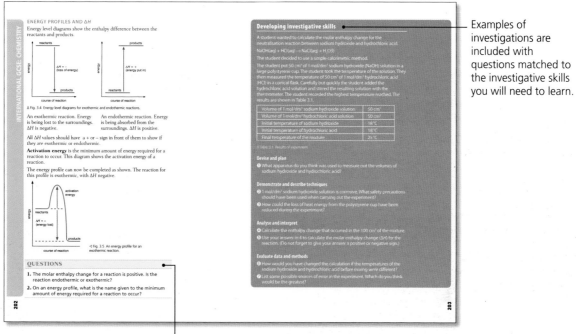

Examples of investigations are included with questions matched to the investigative skills you will need to learn.

Questions to check understanding.

Getting the best from the book *continued*

Science in context boxes put the ideas you are learning into a historical or modern context.

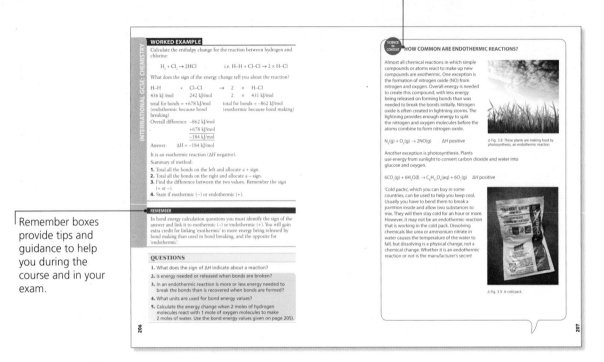

Remember boxes provide tips and guidance to help you during the course and in your exam.

Extension boxes extend learning further.

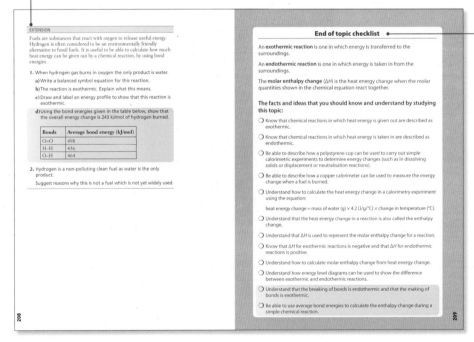

A full checklist of all the information you need to cover the complete specification requirements for each topic.

End of topic questions where you need to apply the knowledge and understanding you have learned in the topic to answer the questions.

The blue side panels and background shading indicate content for Chemistry International GCSE students only.

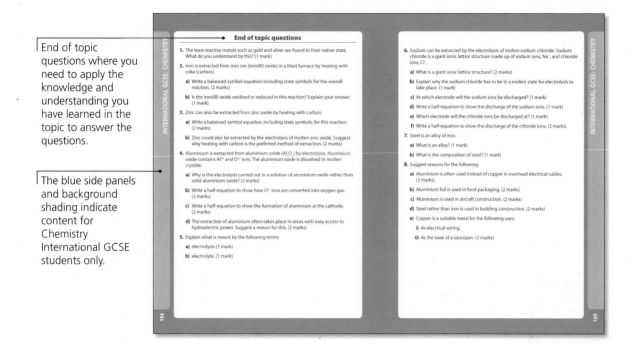

The first question is a student sample with examiners' comments to show best practice.

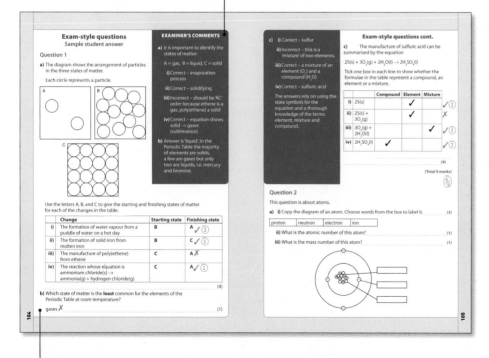

Each section includes exam-style questions to help you get the best results.

This section provides the foundations that the rest of your course is built on. You may have covered some of the topics in your previous work in chemistry but it is important to have a secure knowledge of the key principles before seeing how these can be applied across all the other sections.

Initially you will look at the structure of an atom and why the atoms of different elements have different properties. You will look at the different ways that atoms of elements join together when they form compounds and how the method of combination will determine the properties of the compound formed. You will develop your skills in writing word and symbolic equations. As well as being able to use an equation to work out *what* the products of a reaction will be, you will be able to calculate *how much* of the product can be made in the reaction. These quantitative aspects of chemistry are crucially important in the chemical industry.

STARTING POINTS

1. What is an atom?

2. Do you know the names of any of the particles that are found in an atom?

3. What name is given to a particle formed when two atoms combine together?

4. You will be learning more about the states of matter. What are these states?

5. One type of chemical bonding you will study is called ionic bonding. What is an ion?

6. Diamond and graphite are both covalent substances. They contain the same atoms but have very different structures and properties. What do you know about the properties of diamond and graphite?

SECTION CONTENTS

a) States of matter

b) Elements, compounds and mixtures

c) Atomic structure

d) The Periodic Table

e) Chemical formulae, equations and calculations

f) Ionic bonding

g) Covalent bonding

h) Metallic bonding

i) Electrolysis

j) Exam-style questions

1
Principles of chemistry

△ Diamond and graphite are both forms of carbon but have quite different properties.

△ Fig. 1.1 Water in all its states of matter.

States of matter

INTRODUCTION

Nearly all substances may be classified as solid, liquid or gas – the states of matter. Each one has a state symbol: (s), (l) and (g). The **kinetic theory** of matter is based on the idea that all substances are made up of extremely tiny particles. The particles in these three states are arranged differently and have different types of movement and different energies. In many cases, matter changes into different states quite easily. The names of many of these processes are in everyday use, such as melting and condensing. Using simple models of the particles in solids, liquids and gases can help to explain what happens when a substance changes state. In addition, ideas on solubility will be explored and solubility curves introduced.

KNOWLEDGE CHECK

✓ Be able to classify substances as solid, liquid or gas.
✓ Be familiar with some of the simple properties of solids, liquids and gases.
✓ Know that all substances are made up of particles.
✓ Know that some solids are soluble in a liquid and form a solution; other solids do not dissolve, they are insoluble.

LEARNING OBJECTIVES

✓ Understand the three states of matter in terms of the arrangement, movement and energy of the particles.
✓ Know the processes through which solids, liquids and gases can be converted from one to the other and back again.
✓ Be able to explain how the arrangement, movement and energy of the particles change in the interconversions between the three states of matter.
✓ Be able to explain the results of experiments involving the dilution of coloured solutions and diffusion of gases.
✓ Know what is meant by solvent, solute, solution, saturated solution and solubility.
✓ Know what is meant by the term solubility in the units of g per 100 g of solvent.
✓ Understand how to plot and interpret solubility curves.
✓ Be able to investigate the solubility of a solid in water at a specific temperature.

◁ Fig. 1.2 Water covers nearly four-fifths of the Earth's surface. In the area of this iceberg all three states of matter exist together: solid water (the ice) is floating in liquid water (the ocean), and the surrounding air contains water vapour (clouds).

HOW DO SOLIDS, LIQUIDS AND GASES DIFFER?

The three states of matter each have different properties, depending on how strongly the particles are held together.

- **Solids** have a fixed volume and shape.
- **Liquids** have a fixed volume but no definite shape. They take up the shape of the container in which they are held.
- **Gases** have no fixed volume or shape. They spread out to fill whatever container or space they are in.

Substances don't always exist in the same state; depending on the physical conditions, they change from one state to another.

Some substances can exist in all three states in the natural world. A good example of this is water.

QUESTIONS

1. What is the state symbol for a liquid?

2. Which is the only state of matter that has a fixed shape?

3. In what ways does fine sand behave like a liquid?

WHY DO SOLIDS, LIQUIDS AND GASES BEHAVE DIFFERENTLY?

The behaviour of solids, liquids and gases can be explained if we think of all matter as being made up of very small particles that are in constant motion. This idea has been summarised in the kinetic theory of matter.

In solids, the particles are held tightly together in a fixed position, so solids have a definite shape. However, the particles are vibrating about their fixed positions because they have energy.

In liquids, the particles are held tightly together but have enough energy to move around. Liquids have no definite shape and will take on the shape of the container they are in.

In gases, the particles are further apart and have enough energy to move apart from each other and are constantly moving. Gas particles can spread apart to fill the container they are in.

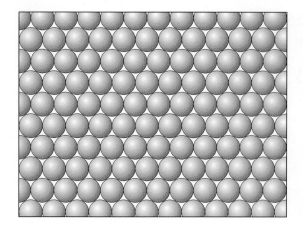

△ Fig. 1.3 Particles in a solid.

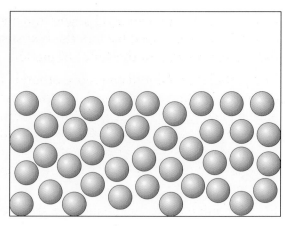

△ Fig. 1.4 Particles in a liquid.

Gases can be compressed to form liquids by using high pressure and cooling.

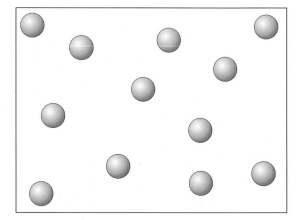

△ Fig. 1.5 Particles in a gas.

HOW DO SUBSTANCES CHANGE FROM ONE STATE TO ANOTHER?

To change solids into liquids and then into gases, they must be heated. Heating provides the particles with enough energy to overcome the forces holding them together.

To change gases into liquids and then into solids involves cooling, removing thermal energy. This makes the particles move closer together as they change from gas to liquid and bond together as the liquid becomes a solid.

The temperatures at which one state changes to another have specific names:

Name of temperature	Change of state
melting point	solid to liquid
boiling point	liquid to gas
freezing point	liquid to solid
condensation point	gas to liquid

△ Table 1.1 Changes of state.

One state of matter can be changed into another state, and in some cases it can be changed back to the first state. Scientists refer to this as the interconversion of the states of matter.

The particles in a liquid can move around. They have different energies, so some are moving faster than others. The faster particles have enough energy to escape from the surface of the liquid and change into gas particles, also called **vapour** particles. This process is **evaporation**. The rate of evaporation increases with temperature, because heat gives more particles the energy to be able to escape from the surface.

Fig. 1.6 summarises the changes in states of matter:

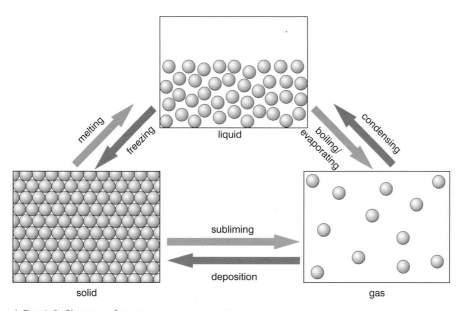

△ Fig. 1.6 Changes of state.

QUESTIONS

1. What type of movement do the particles in a solid have?

2. In which state are the particles held together more strongly: in solid water, liquid water or water vapour?

3. What is the name of the process that occurs when the faster-moving particles in a liquid escape from its surface?

4. What name is given to the temperature at which a solid changes into a liquid?

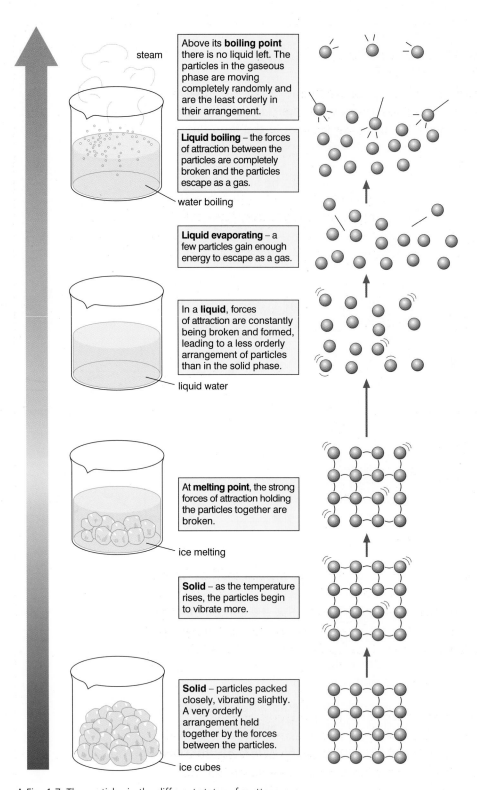

steam

Above its boiling point there is no liquid left. The particles in the gaseous phase are moving completely randomly and are the least orderly in their arrangement.

Liquid boiling – the forces of attraction between the particles are completely broken and the particles escape as a gas.

water boiling

Liquid evaporating – a few particles gain enough energy to escape as a gas.

In a **liquid**, forces of attraction are constantly being broken and formed, leading to a less orderly arrangement of particles than in the solid phase.

liquid water

At **melting point**, the strong forces of attraction holding the particles together are broken.

ice melting

Solid – as the temperature rises, the particles begin to vibrate more.

Solid – particles packed closely, vibrating slightly. A very orderly arrangement held together by the forces between the particles.

ice cubes

Δ Fig. 1.7 The particles in the different states of matter.

THE STATES OF MATTER

There are three states of matter – or are there? To complicate this simple proposition, some substances show the properties of two different states of matter. Some examples are given below.

Liquid crystals

Liquid crystals are commonly used in displays in computers and televisions. Within particular temperature ranges they have the property of a liquid (that is, the particles flow as a liquid) but also of a solid (the particles are arranged in a particular pattern and cannot rotate).

◁ Fig. 1.8 An LCD (liquid crystal display) television.

Superfluids

When some liquids are cooled to very low temperatures, they form a second liquid state described as a superfluid state. Liquid helium at just above absolute zero has infinite fluidity and will 'climb out' of its container when left undisturbed (the liquid at this temperature has zero viscosity). (You may like to look up 'fluidity' and 'viscosity'.)

Plasma

Plasmas or ionised gases can exist at temperatures of several thousand degrees Celsius. An example of plasma is the charged air produced by lightning. Stars like our Sun also produce plasma. Like a gas, a plasma does not have a definite shape or volume but the strong forces between its particles give it unusual properties, such as conducting electricity. Because of this combination of properties, plasma is sometimes called the fourth state of matter.

DIFFUSION EXPERIMENTS

Scientists believe the kinetic theory because of the evidence from simple experiments.

The random mixing and moving of particles in liquids and gases is known as **diffusion**. The examples given below show the effects of diffusion.

DISSOLVING CRYSTALS IN WATER

Fig. 1.9 shows purple crystals of potassium manganate(VII) dissolving in water.

There are no water currents, so only the kinetic theory can explain this. The particles of the crystal gradually move into the water and mix with the water particles. In a similar way, if water is added to a solution of potassium manganate(VII) diffusion will eventually leave a solution of uniform colour throughout.

△ Fig. 1.9 Crystals of potassium manganate(VII) dissolving in water.

MIXING GASES

These photos show a jar of air and a jar of bromine gas. Bromine gas is red-brown to orange-brown and heavier than air. The jar of air has been placed on top of the jar of bromine and the lids removed so the gases can mix (left-hand part of Fig. 1.10).

After about 24 hours (right-hand photo) the bromine gas has spread out throughout both jars. Kinetic theory says that the particles of bromine gas can move around randomly so that they can fill both gas jars. This also occurs with hydrogen and air.

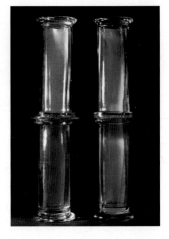

△ Fig. 1.10 Diffusion of bromine.

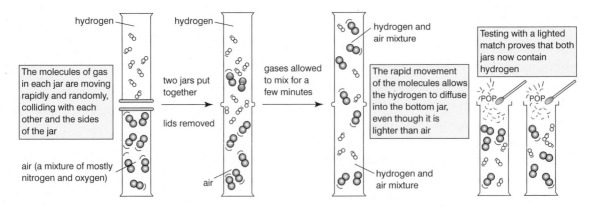

hydrogen

The molecules of gas in each jar are moving rapidly and randomly, colliding with each other and the sides of the jar

air (a mixture of mostly nitrogen and oxygen)

hydrogen

two jars put together

lids removed

air

gases allowed to mix for a few minutes

hydrogen and air mixture

The rapid movement of the molecules allows the hydrogen to diffuse into the bottom jar, even though it is lighter than air

hydrogen and air mixture

Testing with a lighted match proves that both jars now contain hydrogen

POP   POP

△ Fig. 1.11 Demonstration of diffusion with a jar of oxygen and a jar of hydrogen.

Developing investigative skills

Two students set up the experiment shown in the diagram. They carefully clamped the long glass tube horizontally. At the same time, they inserted the cotton wool plugs soaked in the two solutions at each end of the tube and replaced the rubber bungs.

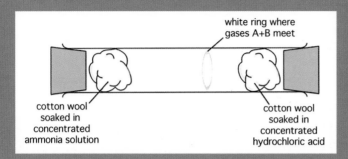

△ Fig. 1.12 Results of experiment.

After about 15 minutes a white ring was seen in the tube.

Note: The white ring was formed where the ammonia gas from the concentrated ammonia solution met the hydrogen chloride gas from the concentrated hydrochloric acid. Together they formed a white substance, ammonium chloride.

Demonstrate and describe techniques

The concentrated ammonia solution is corrosive – it burns and is dangerous to the eyes and dangerous for the environment. It causes severe skin burns and eye damage, and it may cause respiratory irritation. It is very toxic to aquatic organisms. Concentrated hydrochloric acid is corrosive – it burns and its vapour irritates the lungs. It causes severe skin burns and eye damage, and it may cause respiratory irritation.

❶ How should the cotton wool plugs have been handled when putting them into the tube?

❷ What other safety precaution(s) should the two students have used?

Make observations and measurements

❸ Which gas moved the furthest in the 15 minutes before the ring formed?

❹ Approximately how much further did this gas travel compared to the other gas?

Analyse and interpret data

❺ The rate of diffusion of a gas depends on the mass of its particles. What conclusion can you draw about the relative masses of the two gases?

QUESTIONS

1. What is diffusion?

2. Explain how the purple colour of potassium manganate(VII) shown in Fig 1.9 spreads through the water

3. A bottle of perfume is broken at one end of a room. Explain why the perfume can soon be smelled all over the room.

SOLUTIONS AND SOLUBILITY

A **solution** is formed when a solid, known as a **solute**, dissolves in a liquid, known as a **solvent**. The **solubility** of a solid or solute is defined as the number of grams (g) of that solute that dissolves in 100 g of solvent at a particular temperature. When no more solute will dissolve in the solvent at a particular temperature the solution is called a **saturated solution**.

So, for example, if 38 g of sodium chloride dissolves in 100 g of water at 40 °C then the solubility will be 38 g/100 g water. If the solution is saturated at this temperature then no more sodium chloride will dissolve.

The changing solubility of solutes with temperature can be shown as **solubility curves**, as shown in Fig. 1.14. You will see that increasing the temperature of the water has a much bigger impact on the solubility of potassium nitrate than on the solubility of sodium chloride.

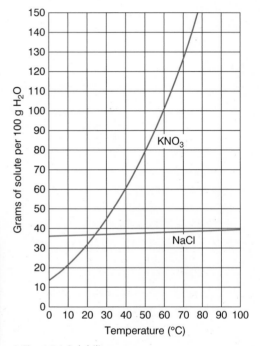

△ Fig. 1.14 Solubility curves.

Developing investigative skills

A student was investigating how the solubility of a solid (sodium bicarbonate) changes with temperature. The student decided to measure the solubility at four different temperatures, 20 °C, 40 °C, 60 °C and 80 °C. The student then selected one of the temperatures and followed the method below:

❶ Weigh 30 g of the solid into a small dry beaker.

❷ Put 100 cm³ of distilled water into 250 cm³ beaker and put it onto the hot plate. Stir the water with a thermometer and adjust the temperature of the hot plate so that the water reaches the temperature you have chosen.

❸ When the required temperature has been reached add a spatula measure of the solid and stir with the thermometer until the solid has dissolved. Then add another spatula of solid and repeat doing this until on adding a spatula of solid the solid does not dissolve completely. It is really important to make sure the temperature remains constant when you are adding the solid.

❹ Reweigh the small beaker and work out how much solid has been added to the water.

The student repeated the method with the other temperatures and recorded the results in the table below:

Temperature (°C)	20	40	60	80
Amount of solid dissolved (g)	10	13	16	20

Analyse and interpret

❶ Plot a graph of the amount of solubility against temperature.

❷ What does the graph indicate about the effect of temperature on the solubility of the solid?

❸ Use your solubility curve to make an accurate estimate of the solubility of the solid at 52 °C.

Evaluate data and methods

❹ What do you think are the main sources of error in this experiment? (Pick the two which you think would have the greatest effect on the accuracy of the results.)

End of topic checklist

Melting is the change of state from solid to liquid.

Boiling and **evaporation** are the change of state from liquid to gas.

Freezing is the change of state from liquid to solid.

Condensation is the change of state from gas to liquid.

Sublimation is the change of state from solid to gas.

Deposition is the change from gas to solid.

Diffusion is the random mixing and moving of particles in liquids and gases.

A **solvent** is a liquid that will dissolve a solid.

A **solute** is a solid that dissolves in a liquid.

A **solution** is formed when a solid dissolves in a liquid.

A **saturated solution** will not dissolve any more solid at that temperature.

Solubility is a measure of how much solid will dissolve in a solvent at a particular temperature.

The facts and ideas that you should know and understand by studying this topic:

○ Be able to use the symbols (s), (l) and (g) to describe the three states of matter.

○ Understand the different arrangement of particles in solids, liquids and gases.

○ Understand the movement and energy of the particles in solids, liquids and gases.

○ Be able to describe how the three states of matter can be interconverted and know the names for these processes.

○ Be able to explain how these interconversions take place by describing changes in the arrangement, movement and energy of the particles involved.

○ Understand how the results of experiments involving the dilution of coloured solutions and diffusion of gases can be explained.

○ Know what is meant by the terms solvent, solute, solution and saturated solution.

○ Know that solubility is measured in units of g per 100 g of solvent.

○ Understand how to plot and interpret solubility curves.

○ Be able to describe how to investigate the solubility of a solid in water at a specific temperature.

End of topic questions

1. In which of the three states of matter are the particles moving at the greatest speed? **(1 mark)**

2. Describe the arrangement and movement of the particles in a liquid. **(2 marks)**

3. In which state of matter do the particles just vibrate about a fixed point? **(1 mark)**

4. Sodium (melting point 98 °C) and aluminium (melting point 660 °C) are both solids at room temperature. From their different melting points, what can you conclude about the forces between the particles in the two metals? **(1 mark)**

5. What is the name of the process involved in each of the following changes of state:

   a) $Fe(s) \rightarrow Fe(l)$? **(1 mark)**

   b) $H_2O(l) \rightarrow H_2O(g)$? **(1 mark)**

   c) $H_2O(g) \rightarrow H_2O(l)$? **(1 mark)**

   d) $H_2O(l) \rightarrow H_2O(s)$? **(1 mark)**

6. Ethanol liquid turns into ethanol vapour at 78 °C. What is the name of this temperature? **(1 mark)**

7. Explain how water in the Earth's polar regions can produce water vapour even when the temperature is very low. **(2 marks)**

8. A student wrote in her exercise book 'The particle arrangement in a liquid is more like the arrangement in a solid than in a gas'. Do you agree with this statement? Explain your reasoning. **(2 marks)**

9. What word is used to describe the rapid mixing and moving of particles in a gas? **(1 mark)**

10. Look at the photographs of the gas jars of air and bromine in Fig. 1.10 on page 16. Explain how bromine gas fills the top gas jar even though it is denser than the air. **(2 marks)**

11. Explain the meaning of the following terms:

    a) saturated solution **(1 mark)**

    b) solubility **(1 mark)**

12. Look at the solubility curves shown in Fig. 1.14 on page 18. Deduce the solubility of potassium nitrate at:

    i) 40 °C **(1 mark)**

    ii) 70 °C **(1 mark)**

△ Fig. 1.15 A model of a molecule.

Elements, compounds and mixtures

INTRODUCTION

The kinetic theory of matter assumes that all substances are made up of small particles. These particles are extremely small and can only be seen using very powerful electron microscopes, but some simple laboratory experiments confirm that they do exist. Most commonly, these particles are atoms and molecules, which make up the elements, compounds and mixtures of many everyday substances. Many of the substances we come across in our lives are mixtures. When necessary, these may be separated into their component parts using a number of simple processes.

KNOWLEDGE CHECK

✓ Be familiar with a simple kinetic theory based on the idea of all matter being made up of particles.
✓ Be able to describe the processes in which solids, liquids and gases can be changed from one to another and back again.

LEARNING OBJECTIVES

✓ Understand how to classify a substance as an element, compound or mixture.
✓ Understand that a pure substance has a fixed melting and boiling point, but that a mixture may melt or boil over a range of temperatures.
✓ Be able to describe how to separate mixtures by simple distillation, fractional distillation, filtration, crystallisation and paper chromatography.
✓ Describe how a chromatogram provides information about the composition of a mixture.
✓ Be able to describe how the calculation of R_f values is used to identify the components of a mixture.
✓ Be able to investigate paper chromatography using inks/food colourings.

ELEMENTS, ATOMS AND COMPOUNDS

All matter is made from **elements**. Elements are substances that cannot be broken down into anything simpler, because they are made up of only one kind of the same small particle. These small particles are called **atoms**.

△ Fig. 1.13 Model of a water – a compound.

Almost always, the atoms in the element combine with other atoms to form **compounds**. For example, the particles in water are molecules containing two hydrogen atoms joined up with one oxygen atom. The formula is therefore H_2O.

ATOMS AND MOLECULES

Atoms and molecules are incredibly small. For example, atoms range in size from 30 to 300 trillionths of a metre! Molecules, which are made up of atoms, are bigger but are still very, very small.

Molecules are the building blocks of most of the substances around us. Even though they are very small, their size and shape are very important and determine how they behave. Here are some examples.

Enzymes are biological catalysts; they speed up chemical reactions. An enzyme molecule is a long chain of amino acids, folded into a ball. There is a dent in the ball into which another molecule can fit. This dent is called the active site of the enzyme. The molecule that fits into the enzyme is called its substrate. Once in the active site the substrate reacts and changes into another molecule which then no longer fits perfectly into the active site.

substrate

enzyme

active site of enzyme

△ Fig. 1.16 An enzyme's substrate fits perfectly into its active site.

Many drugs work by jamming these active sites in enzyme molecules. It is a bit like jamming a lock by inserting a key that is too big for the lock.

MIXTURES AND COMPOUNDS

A **mixture** contains two or more substances. No chemical reaction takes place when the mixture is formed. The individual substances in a mixture can be separated again quite easily using physical methods.

Chemical **compounds** are formed when atoms join together in a chemical reaction. The properties of a compound are very different from the properties of the components that were used to form the compound. The components can only be separated by another chemical reaction.

A key difference between a mixture and a pure substance is that a mixture may melt or boil over a range of temperatures, whereas a pure substance will melt at a fixed melting or boiling point. This fact can be used to distinguish between a mixture and a pure substance. If a substance has a sharp (single temperature) melting or boiling point, it is very likely to be a pure substance.

HOW CAN MIXTURES BE SEPARATED?

Distillation
Distillation is a way of separating a mixture of a solid and a liquid from a solution. For example, pure water can be obtained from salt water by distillation.

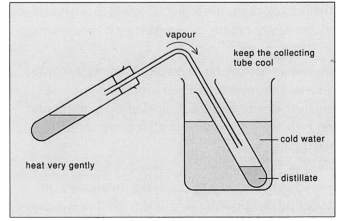
△ Fig. 1.18 Simple method of distillation.

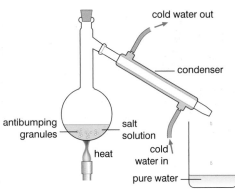
△ Fig. 1.17 Distillation of salt water.

Fractional distillation
A fractionating column separates a mixture of liquids into different fractions or separate substances. It relies on the fact that liquids boil at different temperatures.

Fractional distillation is used for oil refining.

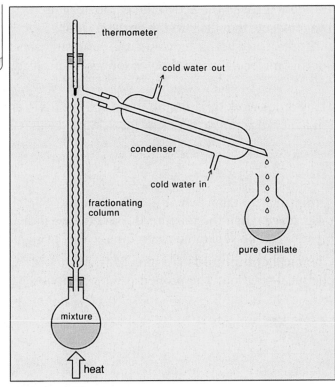
△ Fig 1.19 Using a fractionating column.

Filtration

Filtration separates a solid from a liquid. For example, coffee grounds can be separated from the coffee by filtering through a filter paper. The filter paper is like a sieve: it has tiny holes that allow the coffee to pass through but leave the ground coffee behind.

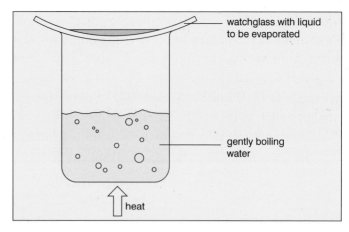

△ Fig. 1.20 Slow method of evaporation.

Crystallisation

If a solid is dissolved in water, it can be recovered by evaporation or crystallisation. As the water evaporates from the mixture, the solid will crystallise.

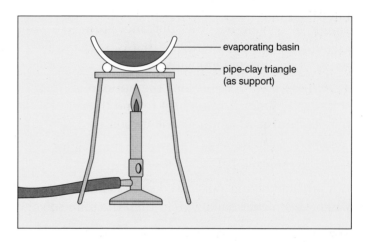

△ Fig. 1.21 Quick method of evaporation.

Paper chromatography

Chromatography can be used to separate a mixture of several solids that are soluble. It is often used to separate coloured substances such as inks or dyes (in Greek *chroma* means colour).

To separate the different coloured dyes in ink, a spot of ink can be placed near the bottom of some filter paper. The filter paper can then be suspended so that it dips into some water in a beaker.

As the solvent spreads through the paper, the dyes are carried in the solvent and they begin to separate. This happens because the different dyes have different solubilities, and so they are carried at different speeds along the paper.

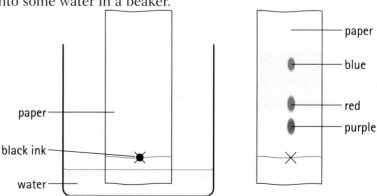

△ Fig. 1.22 Paper chromatography.

Retention factors

Substances can also be identified using chromatography by measuring their **retention factor** on the filter paper. The retention factor (R_f) for a particular substance compares the distance the substance has travelled up the filter paper with the distance travelled by the **solvent**. The retention factor can be calculated using the following formula:

$$R_f = \frac{\text{Distance moved by a substance from the baseline}}{\text{Distance moved by the solvent from the baseline}}$$

As the solvent will always travel further than the substance, R_f values will always be less than 1.

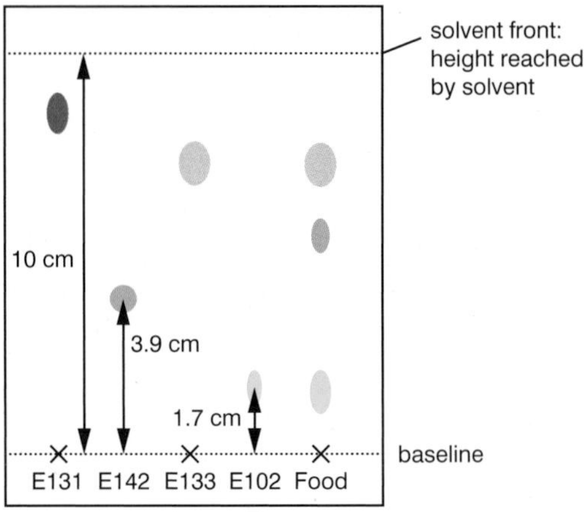

The purity of solids and liquids

It is very important that manufactured foods and drugs contain only the substances the manufacturers want in them – that is, they must not contain any contaminants.

The simplest way of checking the purity of solids and liquids is using heat to find the temperature at which they melt or boil.

An impure solid will have a lower melting point than the pure solid.

A liquid containing a dissolved solid (solute) will have a higher boiling point than the pure solvent.

△ Fig. 1.23 The R_f value for the food additive E102 is 0.17.

The best examples to use to remember these facts are water and ice:

- Pure water boils at $100\,^\circ\text{C}$ – salted water for cooking vegetables boils at about $102\,^\circ\text{C}$.
- Pure ice melts at $0\,^\circ\text{C}$ – ice with salt added to it melts at about $-4\,^\circ\text{C}$.

QUESTIONS

1. What is the advantage of distilling a mixture using a condenser rather than using the simple method shown on page 24?

2. To separate two liquids by fractional distillation they must have different:

   a) melting points **b)** boiling points **c)** colours **d)** viscosities.

3. Which of the three dyes (blue, red and purple) shown in Fig. 1.22 is the most soluble in water?

4. Look at the diagram in Fig. 1.23. Explain why the retention factor (R_f) for the food additive E102 is 0.17.

5. A sample of water contains some dissolved impurities. What would you expect the boiling point of the sample to be?

Developing investigative skills

A student wanted to compare the dyes used in four different samples of black ink using chromatography. The student drew a pencil line near to the bottom of a piece of filter paper and carefully added a dot of each ink on the line and marked them in pencil A, B, C and D. The student then put the filter paper into a beaker containing a small amount of water. When the water had soaked nearly to the top of the filter paper the student removed it from the water and left it to dry. The chromatogram produced is shown in Fig. 1.24.

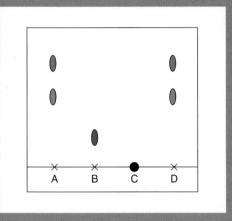

△ Fig 1.24 Results of experiment.

Devise and plan investigations
❶ Why did the student draw the line on the filter paper in pencil?
❷ Why is it important that the level of water in the beaker is below the pencil line?

Analyse and interpret data
❸ Which two inks appear to be the same? Explain your reasoning.
❹ One of the inks has not moved up the filter paper. Suggest a reason for this.
❺ Estimate the retention factor (R_f) for the red dye in ink D.

End of topic checklist

Atoms are the building blocks of all materials.

Molecules are formed when atoms combine together.

An **element** is a substance that cannot be broken down into other substances by any chemical change.

A **compound** is a pure substance formed when elements react together.

Distillation is the process of separating a dissolved solid from its solvent/ liquid.

Chromatography is the process of separating different components of a mixture when they are dissolved in the same solvent.

The **retention factor** (R_f) is used in chromatography. It is the distance travelled by a substance from the baseline divided by the distance travelled by the solvent from the baseline.

The facts and ideas that you should know and understand by studying this topic:

◯ Understand that a pure substance has a fixed melting and boiling point, but that a mixture may melt or boil over a range of temperatures.

◯ Understand the differences between elements, compounds and mixtures.

◯ Be able to describe techniques for separating mixtures including:

- simple distillation to separate a solid and liquid from a solution
- fractional distillation to separate liquids with different boiling points
- filtration to separate a solid from a liquid
- crystallisation to obtain crystals from a solution
- paper chromatography to separate different soluble solids from a solution.

◯ Be able to explain how information from chromatograms can be used to identify the composition of a mixture.

◯ Understand how to use the calculation of R_f values to identify the components of a mixture.

End of topic questions

1. How could measuring the melting point of a solid help to decide whether it was a mixture or a compound? **(1 mark)**

2. What is the main difference between a compound and a mixture? **(1 mark)**

3. What process could be used to separate the following mixtures:

   a) The dyes in the ink of a black felt tip pen? **(1 mark)**

   b) Sand from a sand/water mixture? **(1 mark)**

   c) Petrol from a mixture of petrol and diesel? **(1 mark)**

   d) Pure water from salt water (common salt solution)? **(1 mark)**

4. What is a distillate? **(1 mark)**

5. You are trying to separate the dyes in a sample of ink using paper chromatography. You set up the apparatus as shown in Fig. 1.22 on page 25. After 20 minutes the black spot is unchanged and the water has soaked nearly to the top of the filter paper.

   a) Explain the likely cause for the black spot remaining unchanged. **(1 mark)**

   b) What could you change which might lead to a successful separation of the dyes? **(1 mark)**

6. A mixture of water and ethanol can be separated by fractional distillation. Explain why this process is successful and how it can be undertaken in the laboratory. **(4 marks)**

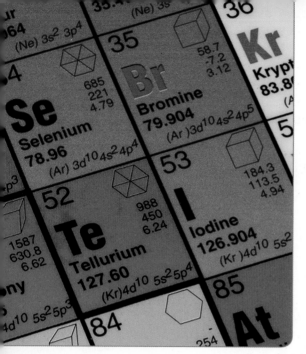

△ Fig. 1.25 The modern Periodic Table.

Atomic structure

INTRODUCTION

The terms atom and molecule were explained in the previous topic. This topic looks in more detail at atoms. Although atoms are the building blocks for all substances, they are made up from even smaller particles – the sub-atomic particles. The types and numbers of these particles in atoms give them their distinctive properties. Because atoms are so small, it is difficult to weigh them, but scientists have developed a relative scale for comparing the mass of different atoms. This is the starting point for calculating the quantities of chemicals needed in a reaction to produce a certain amount of the products.

KNOWLEDGE CHECK

✓ Know that substances are made up of very small particles called atoms.
✓ Understand how diffusion experiments provide evidence for the existence of particles.
✓ Know that chemical compounds contain more than one atom and are often made up of molecules.

LEARNING OBJECTIVES

✓ Know what is meant by the terms atom and molecule.
✓ Know the structure of an atom in terms of the positions, relative masses and relative charges of sub-atomic particles.
✓ Know what is meant by the terms atomic number, mass number, isotopes and relative atomic mass (A_r).
✓ Be able to calculate the relative atomic mass of an element (A_r) from isotopic abundances.

SUB-ATOMIC PARTICLES

The smallest amount of an element that still behaves like that element is an atom. Each element has its own unique type of atom. Atoms are made up of smaller, sub-atomic particles. The three main sub-atomic particles are **protons**, **neutrons** and **electrons**.

These particles are very small and have very little mass. However, it is possible to compare their masses using a relative scale. Their charges may also be compared in a similar way. The proton and neutron have the same mass, and the proton and electron have equal but opposite charges.

Sub-atomic particle	Relative mass	Relative charge
proton	1	+1
neutron	1	0
electron	about $\dfrac{1}{2000}$	−1

△ Table 1.2 Relative masses and charges of sub-atomic particles.

Protons and neutrons are found in the centre of the atom in a cluster called the **nucleus**. The electrons form a series of 'shells' around the nucleus.

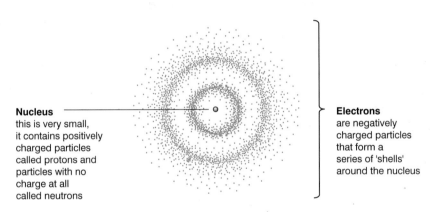

Nucleus
this is very small, it contains positively charged particles called protons and particles with no charge at all called neutrons

Electrons
are negatively charged particles that form a series of 'shells' around the nucleus

△ Fig. 1.26 Structure of an atom.

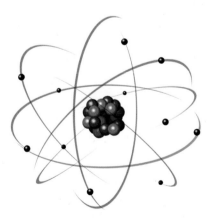

△ Fig. 1.27 Another way of representing the structure of an atom.

ATOMIC NUMBER AND MASS NUMBER

In order to describe the numbers of protons, neutrons and electrons in an atom, scientists use two numbers. These are called the **atomic number** and the **mass number**.

As we shall see in the next topic, atomic numbers are used to arrange the elements in the Periodic Table. The atomic structures of the first ten elements in the Periodic Table are shown in Table 1.3.

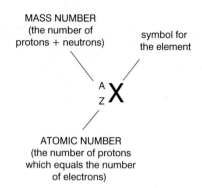

MASS NUMBER (the number of protons + neutrons)

symbol for the element

$$^A_Z X$$

ATOMIC NUMBER (the number of protons which equals the number of electrons)

△ Fig. 1.28 Chemical symbol showing mass number and atomic number.

Hydrogen is the only atom that has no neutrons.

Element	Atomic number	Mass number	Number of protons	Number of neutrons	Number of electrons
hydrogen	1	1	1	0	1
helium	2	4	2	2	2
lithium	3	7	3	4	3
beryllium	4	9	4	5	4
boron	5	10	5	5	5
carbon	6	12	6	6	6
nitrogen	7	14	7	7	7
oxygen	8	16	8	8	8
fluorine	9	19	9	10	9
neon	10	20	10	10	10

△ Table 1.3 Atomic structures of the first ten elements.

QUESTIONS

1. Which sub-atomic particle has the smallest relative mass?

2. Why do atoms have the same number of protons as electrons?

ISOTOPES

Atoms of the same element with the same number of protons and electrons but different numbers of neutrons are called **isotopes**. For example, there are two isotopes of chlorine:

Symbol	Number of neutrons
$^{35}_{17}\text{Cl}$	18
$^{37}_{17}\text{Cl}$	20

△ Table 1.4 Isotopes of chlorine.

Isotopes have the same chemical properties but slightly different physical properties.

 SCIENCE IN CONTEXT **SUB-ATOMIC PARTICLES**

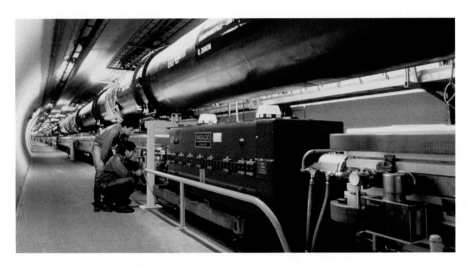

△ Fig. 1.29 The Large Hadron Collider at Cern in Switzerland.

Protons, neutrons and electrons are the sub-atomic particles – it all sounds very straightforward. However, in the past 20 years or so, scientists have discovered a number of other sub-atomic particles: quarks, leptons, muons, neutrinos, bosons and gluons. The properties of some of these other particles have become well known, but there is still much to learn about the others. Finding out about these and possibly other sub-atomic particles is one of the challenges of the 21st century.

To study the smallest known particles, a particle accelerator has been built underground at Cern near Geneva, Switzerland. This giant instrument, called the Large Hadron Collider (LHC) has a circumference of 27 km. It attempts to recreate the conditions that existed just after the 'Big Bang' by colliding beams of particles at very high speed – only about 5 m/s slower than the speed of light. It promises to revolutionise scientific understanding of the nature of atoms. Who knows – school science in 10 or 20 years' time may have changed a lot from your lessons today!

RELATIVE ATOMIC MASS

Atoms are far too light to be weighed. Instead, scientists have developed a **relative atomic mass** scale. The lightest atom, hydrogen, was chosen at first as the unit that all other atoms were weighed against.

On this scale, a carbon atom weighs the same as 12 hydrogen atoms, so carbon's relative atomic mass was given as 12.

Using this relative mass scale you can see, for example, that:

- 1 atom of magnesium has 24 × the mass of 1 atom of hydrogen.
- 1 atom of magnesium has 2 × the mass of 1 atom of carbon.
- 1 atom of copper has 2 × the mass of 1 atom of sulfur.

	Hydrogen	**Carbon**	**Oxygen**	**Magnesium**	**Sulfur**	**Calcium**	**Copper**
Symbol	H	C	O	Mg	S	Ca	Cu
Relative atomic mass	1	12	16	24	32	40	64
Relative size of atom							

△ Table 1.5 Relative atomic masses and sizes of atoms.

Since 1961 the reference point of the relative atomic scale has been carbon-12.

The relative atomic mass, A_r, is the average mass of an atom of an element on a scale in which the mass of one atom of carbon-12 is 12 units. This takes into account the abundance of all existing isotopes of that element.

CALCULATING RELATIVE ATOMIC MASS FROM ISOTOPIC ABUNDANCES

Where an element has different isotopes the relative atomic mass needs to reflect the relative abundances of the isotopes. For example, the element chlorine has two main isotopes and the worked example below shows how the relative atomic mass of chlorine is worked out.

WORKED EXAMPLES

Chlorine has two isotopes, chlorine-35 of 75% abundance and chlorine-37 of 25% abundance. What is the relative atomic mass of chlorine?

Assuming a representative sample of 100 atoms is chosen, 75 of them will each have an atomic mass of 35 and 25 of them will have an atomic mass of 37.

The total atomic mass of the chlorine-35 atoms will be 75×35.

The total atomic mass of the chlorine-37 atoms will be 25×37.

The average atomic mass of each atom (the relative atomic mass) can then be worked out.

The relative atomic mass of chlorine is:

$$\frac{(75 \times 35) + (25 \times 37)}{100} = 35.5$$

The relative atomic mass of chlorine is 35.5 because of the relative abundances of its isotopes.

QUESTIONS

1. What are isotopes?

2. Bromine has two main isotopes: bromine-79 with 51% abundance and bromine-81 with 49% abundance. What is the relative atomic mass of bromine? Give your answer to two decimal places.

End of topic checklist

The **atomic number** is the number of protons (and electrons) in an atom.

The **mass number** is the total number of protons and neutrons in an atom.

The **relative atomic mass** of an atom is the average (mean) mass of an atom on a scale in which the mass of one atom of carbon-12 is 12 units.

Isotopes of an element are atoms with different numbers of neutrons but the same numbers of protons and electrons.

The facts and ideas that you should know and understand by studying this topic:

◯ Know what is meant by the terms atom and molecule.

◯ Understand that atoms are made up of a central nucleus, composed of protons and neutrons, surrounded by electrons arranged in shells.

◯ Know the relative masses and charges of protons, neutrons and electrons.

◯ Be able to calculate the relative atomic mass of an element from the relative abundances of its isotopes.

End of topic questions

1. What is the relative mass of a proton? (1 mark)

2. Explain the meanings of:

   a) atomic number. (1 mark)

   b) mass number. (1 mark)

3. Chlorine has two common isotopes, chlorine-35 and chlorine-37.

   a) What is an isotope? (1 mark)

   b) What are the numbers of protons, neutrons and electrons in each isotope? (2 marks)

4. Copy and complete the table. (4 marks)

Atom	Number of protons	Number of neutrons	Number of electrons
$^{28}_{14}\text{Si}$			
$^{24}_{12}\text{Mg}$			
$^{32}_{16}\text{S}$			
$^{40}_{18}\text{Ar}$			

5. The table below shows information about the structure of six particles (A–F).

In each of the questions i to v, choose one of the six particles A–F. Each letter may be used once, more than once or not at all.

Particle	Protons (positive charge)	Neutrons (neutral)	Electrons (negative charge)
A	8	8	10
B	12	12	10
C	6	6	6
D	8	10	10
E	6	8	6
F	11	12	11

Choose a particle that:

   i) has a mass number of 12 (1 mark)

   ii) has the highest mass number (1 mark)

   iii) has no overall charge (1 mark)

   iv) has an overall positive charge (1 mark)

   v) is the same element as particle E. (1 mark)

The Periodic Table

INTRODUCTION

With over 100 different elements in existence, it's very important to have some way of ordering them. The **Periodic Table** puts elements with similar properties into columns, with a gradual change in properties moving from left to right along the rows. This topic looks at some of the basic features of the Periodic Table. Later topics will look in more detail at particular elements.

Fig. 1.30 This ordering of elements was first published in 1871 by the Russian chemist Dmitri Mendeleev.

KNOWLEDGE CHECK

✓ Understand that all matter is made up of elements.
✓ Know that the atomic number of an element gives the number of protons (and electrons) in an atom of the element.
✓ Know that electrons are arranged in shells around the nucleus of the atom.

LEARNING OBJECTIVES

✓ Understand how elements are arranged in the Periodic Table in order of atomic number, and in groups and periods.
✓ Be able to work out the electronic configurations of the first 20 elements from their positions in the Periodic Table.
✓ Be able to to use electrical conductivity and the acid–base character of oxides to classify elements as metals or non-metals.
✓ Know how to identify elements as metals or non-metals using the Periodic Table.
✓ Be able to describe how the electronic configuration of a main group element is related to its position in the Periodic Table.
✓ Be able to explain why elements in the same group of the Periodic Table have similar chemical properties.
✓ Be able to explain why the noble gases (Group 0) are unreactive.

THE ARRANGEMENT OF THE PERIODIC TABLE

As new elements were discovered in the 19th century, chemists tried to organise the known elements into patterns based on the similarities in their properties. The English chemist John Newlands tried to classify elements according to their properties. The modern Periodic Table is generally thought to have developed from work done by Mendeleev. When the structure of the atom was better known, elements were arranged in order of increasing atomic number, and then the patterns started to make more sense.

HOW ARE ELEMENTS CLASSIFIED IN THE MODERN PERIODIC TABLE?

More than 100 elements have now been identified, and each element has its own properties and reactions. In the Periodic Table, elements with similar properties and reactions are shown close together.

The Periodic Table arranges the elements in order of increasing atomic number. They are then arranged in **periods** and **groups**.

Groups	1	2											3	4	5	6	7	0
Periods																		
1							H hydrogen 1											He helium 2
2	Li lithium 3	Be beryllium 4				transition metals							B boron 5	C carbon 6	N nitrogen 7	O oxygen 8	F fluorine 9	Ne neon 10
3	Na sodium 11	Mg magnesium 12											Al aluminium 13	Si silicon 14	P phosphorus 15	S sulfur 16	Cl chlorine 17	Ar argon 18
4	K potassium 19	Ca calcium 20	Sc scandium 21	Ti titanium 22	V vanadium 23	Cr chromium 24	Mn manganese 25	Fe iron 26	Co cobalt 27	Ni nickel 28	Cu copper 29	Zn zinc 30	Ga gallium 31	Ge germanium 32	As arsenic 33	Se selenium 34	Br bromine 35	Kr krypton 36
5	Rb rubidium 37	Sr strontium 38	Y yttrium 39	Zr zirconium 40	Nb niobium 41	Mo molybdenum 42	Tc technetium 43	Ru ruthenium 44	Rh rhodium 45	Pd palladium 46	Ag silver 47	Cd cadmium 48	In indium 49	Sn tin 50	Sb antimony 51	Te tellurium 52	I iodine 53	Xe xenon 54
6	Cs caesium 55	Ba barium 56	La lanthanum 57	Hf hafnium 72	Ta tantalum 73	W tungsten 74	Re rhenium 75	Os osmium 76	Ir iridium 77	Pt platinum 78	Au gold 79	Hg mercury 80	Tl thallium 81	Pb lead 82	Bi bismuth 83	Po polonium 84	At astatine 85	Rn radon 86

metal | non metal | transition metal | metalloid

△ Fig. 1.31 The Periodic Table.

Periods

Horizontal rows of elements are arranged in increasing atomic number from left to right. Rows correspond to periods, which are numbered from 1 to 7.

Groups

Vertical columns contain elements with the atomic number increasing down the column. They are numbered from 1 to 7 and 0 (Group 0 is often referred to as Group 8).

Groups are sometimes referred to as 'families' of elements – the alkali metals (Group 1), the alkaline earth metals (Group 2) and the halogens (Group 7).

REMEMBER

• It is important to understand the relationship between group number, number of outer electrons, and metallic and non-metallic character across periods.

1. Find the element calcium in the Periodic Table. Answer these questions about calcium:

   a) What is its atomic number?

   b) What information does the atomic number give about the structure of a calcium atom?

   c) Which group of the Periodic Table is calcium in?

   d) Which period of the Periodic Table is calcium in?

   e) Is calcium a metal or a non-metal?

2. What is the family name for the Group 7 elements?

3. Are the Group 7 elements metals or non-metals?

METALS AND NON-METALS

Most elements can be classified as either metals or non-metals. In the Periodic Table, the metals are arranged on the left and in the middle, and the non-metals are on the right.

Metalloid elements are between metals and non-metals. They have some properties of metals and some of non-metals. Examples of metalloids are antimony (Sb) and germanium (Ge).

Metals and non-metals have quite different physical and chemical properties.

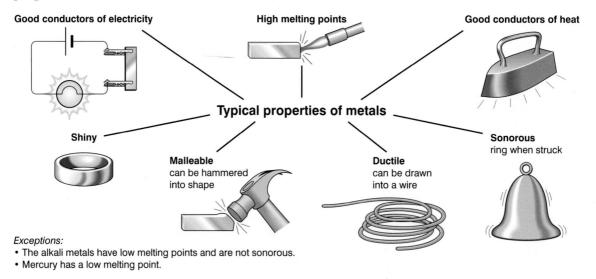

Exceptions:
- The alkali metals have low melting points and are not sonorous.
- Mercury has a low melting point.

Δ Fig. 1.32 Properties of metals.

△ Fig. 1.33 Metals: chromium, manganese, iron, cobalt, nickel, copper and zinc.

Metal oxides form **basic oxides**. Basic oxides, which do not dissolve in water, will react with acids to form chemicals called **salts** (for more detail see page 168). Metal oxides, which dissolve in water, form **alkalis** (for more detail see page 110).

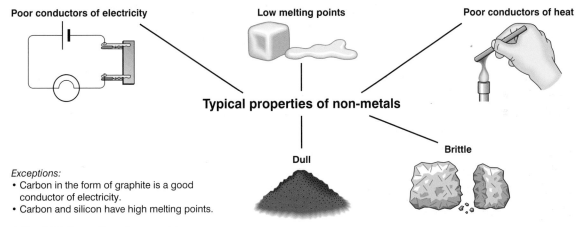

Poor conductors of electricity

Low melting points

Poor conductors of heat

Typical properties of non-metals

Dull

Brittle

Exceptions:
- Carbon in the form of graphite is a good conductor of electricity.
- Carbon and silicon have high melting points.

△ Fig. 1.34 Properties of non-metals.

△ Fig. 1.35 Non-metals from left: silicon, chlorine, sulfur.

Non-metal oxides that dissolve in water typically form **acidic oxides** or **acids**. Acidic oxides react with alkalis to form salts.

1. What sort of element is a metalloid?

2. Metals are often ductile. What does this mean?

3. Metals are usually malleable. What does this mean?

4. Which non-metal is an exception to the rule and does conduct electricity?

5. What sort of oxides do non-metals usually form?

The electrons are arranged in **shells** around the nucleus. These do not all contain the same number of electrons – the shell nearest to the nucleus can only take two electrons, whereas the next one out from the nucleus can take eight.

Electron shell	Maximum number of electrons
1	2
2	8
3	8 (initially with up to 18 after element 20)

△ Table 1.6 Maximum number of electrons in a shell.

Oxygen has an atomic number of 8, so it has eight electrons. Of these, two are in the first shell and six are in the second shell. This arrangement is written 2, 6. A phosphorus atom with an atomic number of 15 has 15 electrons, arranged 2, 8, 5.

The atomic structure of an atom can be shown simply in a diagram.

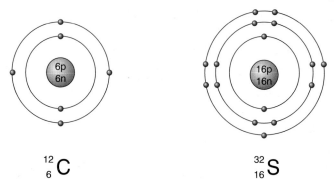

$${}^{12}_{6}C$$ $${}^{32}_{16}S$$

△ Fig. 1.36 Atomic diagrams for carbon and sulfur showing the number of protons and neutrons and the electron arrangements.

The arrangement of electrons in an atom is called its **electronic configuration**.

There are over 100 different elements. They may be arranged in the Periodic Table according to their chemical and physical properties.

The chemical properties of elements depend on the arrangement of electrons in the atoms. The electronic configuration of the first 20 elements is shown in Table 1.7.

Element	Symbol	Atomic number	Electron number	Electronic configuration
hydrogen	H	1	1	1
helium	He	2	2	2
lithium	Li	3	3	2, 1
beryllium	Be	4	4	2, 2
boron	B	5	5	2, 3
carbon	C	6	6	2, 4
nitrogen	N	7	7	2, 5
oxygen	O	8	8	2, 6
fluorine	F	9	9	2, 7
neon	Ne	10	10	2, 8
sodium	Na	11	11	2, 8, 1
magnesium	Mg	12	12	2, 8, 2
aluminium	Al	13	13	2, 8, 3
silicon	Si	14	14	2, 8, 4
phosphorus	P	15	15	2, 8, 5
sulfur	S	16	16	2, 8, 6
chlorine	Cl	17	17	2, 8, 7
argon	Ar	18	18	2, 8, 8
potassium	K	19	19	2, 8, 8, 1
calcium	Ca	20	20	2, 8, 8, 2

△ Table 1.7 Electronic configuration of first 20 elements.

Elements that have similar electronic configurations have similar chemical properties.

Lithium (2, 1), sodium (2, 8, 1) and potassium (2, 8, 8, 1) all have one electron in their outer shell. These are all highly reactive metals. They are called Group 1 elements in the Periodic Table.

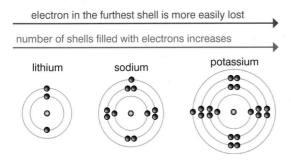

△ Fig. 1.37 Electronic configuration of lithium, sodium and potassium

Fluorine (2, 7), chlorine (2, 8, 7), bromine (2, 8, 18, 7) and iodine (2, 8, 18, 18, 7) all have seven electrons in their outer shell. These elements are all highly reactive non-metals. They are called Group 7 elements, or halogens.

Similarly, all the elements in Group 3 of the Periodic Table have three electrons in their outer electron shell.

The elements helium (2), neon (2, 8), argon (2, 8, 8), krypton (2, 8, 18, 8) and xenon (2, 8, 18, 18, 8) either have a full outer shell or have eight electrons in their outer shell and the atoms therefore do not lose or gain electrons easily. This means that these gases are unreactive. They are called **noble gases**.

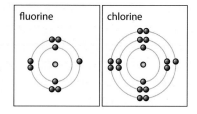

△ Fig. 1.38 Electronic configuration of fluorine and chlorine.

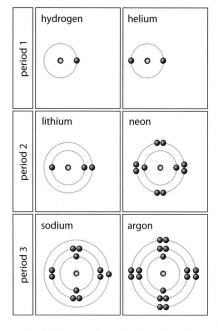

△ Fig. 1.39 Electronic configuration of helium, neon and argon.

△ Fig. 1.40 Neon lighting in Hong Kong.

QUESTIONS

1. a) How many electrons does magnesium have in its outer electron shell?

b) Which group of the Periodic Table is magnesium in?

2. Draw atom diagrams for:

a) aluminium

b) calcium.

3. Why are the noble gases unreactive?

ELECTRONIC CONFIGURATION AND THE PERIODIC TABLE

Elements with the same number of electrons in their outer shells have similar chemical properties. The relationship between the group number and the number of electrons in the outer electron shell is shown in Table 1.8.

Group number	1	2	3	4	5	6	7	0 (8)
Electrons in the outer electron shell	1	2	3	4	5	6	7	2 or 8

△ Table 1.8 Relationship between group number and number of electrons in outer shell.

There is also a link between the period number and the number of electron shells. As the period number increases so too will the number of electron shells. For example, helium is in the first period of the Periodic Table and it has one shell of electrons. In contrast, potassium is in the fourth period of the Periodic Table and has four shells of electrons.

REACTIVITIES OF ELEMENTS

Going from the top to the bottom of a group in the Periodic Table, metals become more reactive but non-metals become less reactive. These trends in reactivity will be explained in the topics on Group 1 and 7 elements (see later).

Group 0 elements, known as the noble gases, are very unreactive. They already have full outer electron shells or eight electrons in the outer shell and so rarely react with other elements to form compounds.

△ Fig. 1.41 Group 1 elements (metals) become more reactive further down the group.

QUESTIONS

1. How many electrons does aluminium have in its outer shell?

2. Which is the most reactive element in Group 7?

3. Which is the most reactive element in Group 2?

△ Fig. 1.42 The Group 7 elements (non-metals) become more reactive further up the group.

SCIENCE IN CONTEXT

THE FIRST PERIODIC TABLE

In 1869 the Russian chemist Dmitri Mendeleev published his work on the Periodic Table. It included the 66 known elements. Interestingly, Mendeleev left gaps in his arrangement when the next element in his order did not seem to fit. He predicted that there should be elements in the gaps but that they had yet to be discovered. One such element is gallium (discovered in 1875), which Mendeleev predicted would be between aluminium and indium.

By June 2011 there were 118 known elements but only 91 of these were naturally-occurring – the others had been made artificially. Some of these artificial elements can be used, for example the element americium (Am, atomic number 95), is used in smoke detectors.

End of topic checklist

A **group** is a column of elements in the Periodic Table.

A **period** is a row of elements in the Periodic Table.

A **basic oxide** is the oxide of a metal.

An **acidic oxide** is the oxide of a non-metal.

An **electronic configuration** provides information on the number and arrangement of electrons.

The facts and ideas that you should know and understand by studying this topic:

○ Know that elements are arranged in the Periodic Table in order of atomic number.

○ Understand how elements are arranged in the Periodic Table in groups and periods.

○ Know that metals are found on the left-hand side and in the middle of the Periodic Table.

○ Know that non-metals are found on the right-hand side of the Periodic Table.

○ Be able to deduce the electronic configurations of the first 20 elements in the Periodic Table.

○ Know that the number given to a group of elements in the Periodic Table is the same as the number of electrons in the outer electron shell of the atoms in the group.

○ Know that metals conduct electricity and form basic oxides.

○ Know that non-metals are generally non-conductors of electricity and usually form acidic oxides.

○ Understand that the chemical properties of an element are largely governed by the number of electrons in its outermost electron shell and so elements in the same group have similar chemical properties.

○ Understand that the noble gases in Group 0 are very unreactive because they have either full outer shells of electrons or eight electrons in the outer shell.

End of topic questions

1. Look at the diagram representing the Periodic Table. The letters stand for elements.

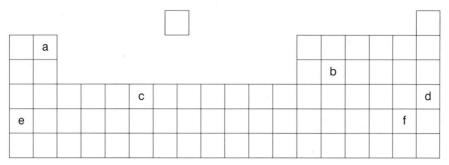

    a) Which element is in Group 4?   (1 mark)

    b) Which element is in the second period?   (1 mark)

    c) Which element is a noble gas?   (1 mark)

    d) Which elements are non-metals?   (1 mark)

2. An element conducts electricity.

    a) Is it likely to be a metal or a non-metal?   (1 mark)

    b) Is it likely to form a basic or acidic oxide? Explain your answer.   (1 mark)

3. Why do elements in the same group have similar chemical properties?   (1 mark)

4. Draw an atom diagram for:

    a) oxygen   (2 marks)

    b) potassium.   (2 marks)

5. Is there a relationship between the group number of the first 20 elements in the Periodic Table and the following:

    a) The number of protons in an atom of the element?   (1 mark)

    b) The number of neutrons in an atom of the element?   (1 mark)

    c) The number of electrons in an atom of the element?   (1 mark)

    d) The number of electrons in the outer electron shell of the element?   (1 mark)

6. In the Periodic Table, what is the trend in reactivity:

    a) Down a group of metals?   (1 mark)

    b) Down a group of non-metals?   (1 mark)

7. Explain why the noble gases in Group 0 are very unreactive.   (2 marks)

△ Fig. 1.43 When this reaction is described as: $S(s) + O_2(g) \rightarrow SO_2(g)$ it is understood by chemists all over the world.

Chemical formulae, equations and calculations

INTRODUCTION

Languages change from country to country, and so too do the spellings of the names of the chemical elements. For example, what we know in English as sodium may be known to chemists in eastern Europe as natrium (this explains the symbol Na). The symbols for the elements and chemical formulae of compounds used to write equations for chemical reactions, however, are the same throughout the world. In this way, chemical equations are the language of the chemist.

KNOWLEDGE CHECK

✓ Understand relative atomic masses.
✓ Know the state symbols for solids, liquids, gases and aqueous solutions.

LEARNING OBJECTIVES

✓ Be able to write word equations and balanced chemical equations which include state symbols.
✓ Be able to calculate relative formula masses.
✓ Know that the mole (mol) is the unit for the amount of a substance.
✓ Be able to use simple calculations involving amount of substance, relative atomic mass (A_r) and relative formula mass (M_r) to work out the masses of reactants and products involved in a chemical reaction.
✓ Be able to calculate percentage yield.
✓ Know what is meant by the terms empirical formula and molecular formula.
✓ Be able to work out an empirical and molecular formula from experimental results.
✓ Be able to determine the formula of a metal oxide by combustion (e.g. magnesium oxide) or by reduction (e.g. copper(II) oxide).
✓ Be able to carry out calculations involving amount of substance, volume and concentration (in mol/dm³) of solution.
✓ Be able to use the molar volume of a gas (24 dm³ and 24 000 cm³ at room temperature and pressure (rtp)) to work out the volumes of gases involved in a chemical reaction.

WRITING CHEMICAL EQUATIONS

In a chemical equation the starting chemicals are called the **reactants** and the finishing chemicals are called the **products**.

Follow these simple rules to write a chemical equation.

1. Write down the word equation.

2. Write down the symbols (for elements) and formulae (for compounds).

3. Balance the equation, to make sure there are the same number of each type of atom on each side of the equation.

4. Include the state symbols solid (s); liquid (l); gas (g); solution in water (aq).

State	State symbol
solid	(s)
liquid	(l)
gas	(g)
solution	(aq)

△ Table 1.9 States and their symbols.

Remember that some elements are **diatomic**. They exist as molecules containing two atoms.

Element	Formula of diatomic molecule
hydrogen	H_2
oxygen	O_2
nitrogen	N_2
chlorine	Cl_2
bromine	Br_2
iodine	I_2

△ Table 1.10 Some diatomic elements.

WORKED EXAMPLES

1. When a lighted splint is put into a test tube of hydrogen the hydrogen burns with a 'pop'. In fact the hydrogen reacts with oxygen in the air (the reactants) to form water (the product). Write the chemical equation for this reaction.

Word equation: hydrogen + oxygen → water

Symbols and formulae: H_2 + O_2 → H_2O

Balance the equation: $2H_2$ + O_2 → $2H_2O$

For every two molecules of hydrogen that react, one molecule of oxygen is needed and two molecules of water are formed.

$$2H_2(g) \qquad + \qquad O_2(g) \qquad \rightarrow \qquad 2H_2O(l)$$

two molecules   one molecule   two molecules

2. What is the equation when sulfur burns in air?

Word equation: sulfur + oxygen → sulfur dioxide

Symbols and formulae: $S(s) + O_2(g) \rightarrow SO_2(g)$

Balance the equation: $S(s) + O_2(g) \rightarrow SO_2(g)$

▷ Fig. 1.44 Methane is burning in the oxygen in the air to form carbon dioxide and water.

BALANCING EQUATIONS

Balancing equations can be quite tricky. Basically it is done by trial and error. However, the golden rule is that *balancing numbers can only be put in front of the formulae*.

For example, to balance the equation for the reaction between methane and oxygen:

	Reactants	Products
Start with the unbalanced equation.	$CH_4 + O_2$	$CO_2 + H_2O$
Count the number of atoms on each side of the equation.	1C ✓, 4H, 2O	1C ✓, 2H, 3O
There is a need to increase the number of H atoms on the products side of the equation. Put a '2' in front of the H_2O.	$CH_4 + O_2$	$CO_2 + 2H_2O$
Count the number of atoms on each side of the equation again.	1C ✓, 4H ✓, 2O	1C ✓, 4H ✓, 4O
There is a need to increase the number of O atoms on the reactant side of the equation. Put a '2' in front of the O_2.	$CH_4 + 2O_2$	$CO_2 + 2H_2O$
Count the atoms on each side of the equation again.	1C ✓, 4H ✓, 4O ✓	1C ✓, 4H ✓, 4O ✓

△ Table 1.11 Balancing the equation for the reaction between methane and oxygen.

No atoms have been created or destroyed in the reaction. The equation is balanced!

$$CH_4(g) + 2O_2(g) \rightarrow CO_2(g) + 2H_2O(l)$$

△ Fig. 1.45 The number of each type of atom is the same on the left and right sides of the equation.

In balancing equations involving **radicals** such as sulfate, hydroxide, nitrate, you can use the same procedure. For example, when lead(II) nitrate solution is mixed with potassium iodide solution, lead(II) iodide and potassium nitrate are produced.

1. **Words:**

   lead(II) nitrate   +   potassium iodide   →   lead(II) iodide   +   potassium nitrate

2. **Symbols:**

   $Pb(NO_3)_2$ + KI → PbI_2 + KNO_3

3. **Balance the nitrates:**

   $Pb(NO_3)_2$ + KI → PbI_2 + $2KNO_3$

4. **Balance the iodides:**

   $Pb(NO_3)_2(aq)$ + $2KI(aq)$ → $PbI_2(s)$ + $2KNO_3(aq)$

△ Fig. 1.46 This reaction occurs simply on mixing the solutions of lead(II) nitrate and potassium iodide. Lead iodide is an insoluble yellow compound.

QUESTIONS

1. Balance the following chemical equations:

a) $Ca(s) + O_2(g) \rightarrow CaO(s)$

b) $H_2S(g) + O_2(g) \rightarrow SO_2(g) + H_2O(l)$

c) $Pb(NO_3)_2(s) \rightarrow PbO(s) + NO_2(g) + O_2(g)$

2. Write balanced symbol equations from the following word equations:

a) sulfur + oxygen → sulfur dioxide

b) magnesium + oxygen → magnesium oxide

c) copper(II) oxide + hydrogen → copper + water

As mentioned earlier, the general method for balancing equations is by trial and error, but it helps if you are systematic: always start on the left-hand side with the reactants.

Sometimes you can balance an equation using fractions. In more advanced study such balanced equations are perfectly acceptable. Getting rid of the fractions is not difficult, though. Look at this example:

WORKED EXAMPLE

Ethane (C_2H_6) is a hydrocarbon fuel and burns in air to form carbon dioxide and water.

Unbalanced equation: $C_2H_6(g) + O_2(g) \rightarrow CO_2(g) + H_2O(l)$

Balancing the carbon and hydrogen atoms gives:

$$C_2H_6(g) + O_2(g) \rightarrow 2CO_2(g) + 3H_2O(l)$$

The equation can then be balanced by putting 3½ in front of the O_2. By doubling every balancing number, the equation is then balanced using whole numbers!

$$2C_2H_6(g) + 7O_2(g) \rightarrow 4CO_2(g) + 6H_2O(l)$$

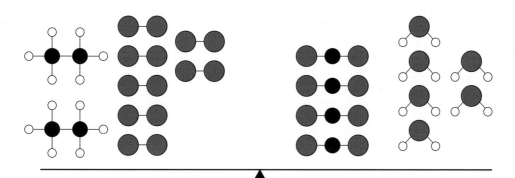

△ Fig. 1.47 Balancing the equation for burning ethane in air.

EXTENSION

During your course you will become familiar with balancing equations and become much quicker at doing it. Try balancing the equations below. The third is the chemical reaction often used for making chlorine gas in the laboratory.

1. Balance the following equations:

a) $C_5H_{10}(g) + O_2(g) \rightarrow CO_2(g) + H_2O(l)$

b) $Fe_2O_3(s) + CO(g) \rightarrow Fe(s) + CO_2(g)$

c) $KMnO_4(s) + HCl(aq) \rightarrow KCl(aq) + MnCl_2(aq) + H_2O(l) + Cl_2(g)$

RELATIVE FORMULA MASSES, M_r

The relative formula mass (M_r) is the total relative mass of the atoms included in the formula. In many cases this is also the molecular formula mass, but this term only applies to substances made of molecules and not ions. A relative formula mass (M_r) can be worked out from the relative atomic masses of the atoms in the formula.

The relative formula mass of a molecule can be worked out by simply adding up the relative atomic masses of the atoms in the molecule. For example:

Water, H_2O (H = 1, O = 16)

The relative formula mass (M_r) = 1 + 1 + 16 = 18

(Note: The subscript $_2$ only applies to the hydrogen atom.)

Carbon dioxide, CO_2 (C = 12, O = 16)

$M_r = 12 + 16 + 16 = 44$

(Note: The subscript $_2$ only applies to the oxygen atom.)

A similar approach can be used for any formula, including ionic formulae (see page 72).

WORKED EXAMPLES

1. Sodium chloride, NaCl (A_r: Na = 23, Cl = 35.5)
The relative formula mass (M_r) = 23 + 35.5 = 58.5
Potassium nitrate, KNO_3 (A_r: K = 39, N = 14, O = 16)
$M_r = 39 + 14 + 16 + 16 + 16 = 101$
(Note: The subscript $_3$ only applies to the oxygen atoms.)

2. Calcium hydroxide, $Ca(OH)_2$ (A_r: Ca = 40, O = 16, H = 1)
$M_r = 40 + (16 + 1)_2 = 40 + 34 = 74$
(Note: The subscript $_2$ applies to everything inside the bracket.)

3. Magnesium nitrate, $Mg(NO_3)_2$ (A_r: Mg = 24, N = 14, O = 16)
$M_r = 24 + (14 + 16 + 16 + 16)_2 = 24 + (62)_2 = 24 + 124 = 148$

QUESTIONS

1. What is the formula mass of methane, CH_4? (A_r: H = 1, C = 12)

2. What is the formula mass of ethanol, C_2H_5OH? (A_r: H = 1, C = 12, O = 16)

3. What is the formula mass of ozone, O_3? (A_r: O = 16)

MOLES

A mole (mol) is the unit for the amount of a substance. One mole of substance is approximately 6×10^{23} particles. This number is called the Avogadro number or constant.

For example, you can have a mole of atoms, a mole of molecules or a mole of electrons. A mole of atoms is 6×10^{23} atoms.

The relative atomic mass of an element tells you the mass of a mole of atoms of that element. So, for example, a mole of carbon atoms has a mass of 12 g.

The **relative formula mass** tells you the mass of a mole of that substance. For example, a mole of sodium chloride, NaCl, has a relative formula mass of 58.5, so 1 mole of sodium chloride has a mass of 58.5 g.

△ Fig. 1.48 Moles vary in mass and volume. What do these samples all have in common?

SCIENCE IN CONTEXT

AMEDEO AVOGADRO

Amedeo Avogadro was born in Italy in 1776. At age 20 he graduated from university with a degree in religious law. However, his real interests were science and mathematics, and this was where he started to devote his university research. At the beginning of the 19th century scientists across the world were trying to work out the relative masses of different atoms. There was much confusion and disagreement about the existence of other particles which we now call molecules. Against this background of confusion in 1811 Avogadro published a research paper on 'determining the relative masses of elementary molecules'. It was in this paper that he proposed his hypothesis:

"All gases under the same conditions of temperature and pressure contain the same number of molecules."

△ Fig. 1.49 Amedeo Avogadro.

The scientific community paid little attention to his work at the time. There were doubts about the existence of molecules and conflicting evidence coming from the research of more well-known and respected scientists. In fact, it was not until 4 years after his death in 1860 that his idea was accepted following work done by a highly respected scientist called Cannizzaro. So Avogadro died without ever knowing the importance of his contribution to the development of science. It was only in 1911 that Avogadro's contribution to chemistry was formally recognised when, in honour of his work, the number of molecules in one mole of a substance was called the Avogadro number, equal to approximately 6×10^{23}.

How many other great advances or discoveries in science have not been recognised during the lifetime of the scientist concerned?

LINKING REACTANTS AND PRODUCTS

Chemical equations allow quantities of reactants and products to be linked together. They tell you how much of the products you can expect to make from a fixed amount of reactants.

In a balanced equation the numbers in front of each symbol or formula indicate the number of moles represented. The number of moles can then be converted into a mass in grams.

For example, when magnesium ($A_r = 24$) reacts with oxygen ($A_r = 16$):

Write down the balanced equation: $2Mg(s) + O_2(g) \rightarrow 2MgO(s)$

Write down the number of moles: $\quad 2 \quad + 1 \quad \rightarrow 2$

Convert moles to masses: $\quad\quad 48\text{ g} \quad + 32\text{ g} \rightarrow 80\text{ g}$

So when 48 g of magnesium reacts with 32 g of oxygen, 80 g of magnesium oxide is produced. From this you should be able to work out the mass of magnesium oxide produced from any mass of magnesium.

◁ Fig. 1.50 In an oxidation reaction, magnesium reacts with oxygen, and the reaction can be used in fireworks and flares producing a brilliant white colour.

WORKED EXAMPLES

1. What mass of magnesium oxide can be made from 6 g of magnesium? (A_r: O = 16, Mg = 24)

Equation:	$2Mg(s)$	+	$O_2(g)$	$\rightarrow$	$2MgO(s)$
Moles:	2		1		2
Masses:	48 g		32 g		80 g
÷ 8	6 g				10 g

Therefore 6 g will form 10 g of MgO

Note: In this example, there was no need to work out the mass of oxygen needed. It was assumed that there would be as much as was necessary to convert all the magnesium to magnesium oxide.

2. What mass of ammonia can be made from 56 g of nitrogen? (A_r: H = 1, N = 14)

Equation:	$N_2(g)$	+	$3H_2(g)$	$\rightarrow$	$2NH_3(g)$
Moles:	1		3		2
Masses:	28 g		6 g		34 g
×2	56 g				68 g

Mass of ammonia = 68 g

EXPERIMENTS TO FIND THE FORMULAE OF SIMPLE COMPOUNDS

Finding the formula of magnesium oxide

Magnesium ribbon can be heated in a crucible to make a white powder called magnesium oxide. The magnesium reacts with oxygen from the air. The reaction is called **oxidation** because a substance combines with oxygen.

If you measure the masses of the reactants in this reaction, you can use the relative atomic masses to work out the formula of the compound made.

Here are some typical results when 0.24 g of magnesium is oxidised.

Measurement	Mass in grams
Mass of crucible + lid	30.00
Mass of crucible + lid + magnesium	30.24
Mass of crucible + lid + magnesium oxide	30.40
Mass of magnesium	30.24 − 30.00 = 0.24
Mass of magnesium oxide	30.40 − 30.0 = 0.40
Mass of oxygen	30.40 − 30.24 = 0.16

The result is that 0.24 g of magnesium joins with 0.16 g of oxygen. Therefore 24 g of magnesium would join with 16 g of oxygen. The relative atomic mass of magnesium is 24 and of oxygen is 16. So 1 mole of magnesium atoms joins with 1 mole of oxygen atoms. This means that the formula of magnesium oxide is simply MgO.

▷ Fig. 1.51 Magnesium ribbon is put in a crucible with a lid on it. The crucible is heated until the magnesium is red hot. The lid is lifted very slightly and put back down again to allow oxygen in. This lets the magnesium burn but prevents loss of magnesium oxide.

WORKED EXAMPLE

Copper(II) oxide can be heated in hydrogen to produce copper and water.

In an experiment 12.8 g of copper was produced from 16.0 g of copper(II) oxide.

Relative atomic masses: (A_r: H = 1, O = 16, Cu = 64).

Word equation: copper(II) oxide + hydrogen $\rightarrow$ copper + water

Masses   16.0 g   12.8 g

So the mass of oxygen that was combined with the copper =
16.0 – 12.8 = 3.2 g.

Therefore 32 g of oxygen would be combined with 128 g of copper.

Therefore = $\frac{32}{16}$ = 2 moles of oxygen would be combined with
$\frac{128}{64}$ = 2 moles of copper.

Therefore 1 mole of oxygen would be combined with 1 mole of copper.

Therefore the formula for copper(II) oxide is CuO.

Developing investigative skills

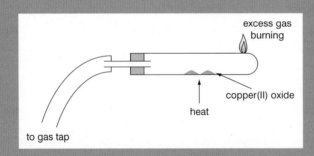

△ Fig. 1.52 Apparatus to find the chemical formula of a sample of copper(II) oxide.

A student set out to find the chemical formula of a sample of copper(II) oxide. First the student weighed an empty combustion tube, then used a spatula to put a full measure of copper(II) oxide near the centre of the tube, being careful not to spill any of the powder near the opening of the tube. The student then reweighed the tube and set up the apparatus as shown in the diagram. (Note: copper(II) oxide is harmful if swallowed. It causes serious eye damage, and skin irritation, and it is dangerous for the environment, as it is very toxic to aquatic organisms.

Holding a lighted splint over the jet, the gas was turned on very slowly until a flame about 1 cm high was burning at the jet. The copper(II) oxide was then heated strongly in a Bunsen flame for about 15 minutes until all the copper (II) oxide had been turned into copper. At this point the tube was allowed to cool, but with the gas still flowing through the tube and being burnt at the jet.

When the centre of the tube was cold, the gas was turned off and the tube and contents reweighed. The results are shown in the table:

Mass of tube + copper(II) oxide	= 24.15 g
Mass of tube + copper	= 23.92 g
Mass of tube	= 23.15 g
Mass of copper(II) oxide	=
Mass of copper	=
Mass of oxygen combined with the copper	=

△ Table 1.12 Results of experiment.
Note: Methane is being used as a safer alternative to hydrogen in this investigation.

Demonstrate and describe techniques

❶ Why was it important not to spill any of the copper(II) oxide near the opening of the tube?

❷ Why was the gas supply turned on very slowly?

❸ Why was the tube allowed to cool before turning off the gas supply?

Make observations and measurements

❹ What colour change would you expect to see as the copper(II) oxide changes into copper?

❺ Use the results to work out the masses of copper(II) oxide, copper and oxygen.

Analyse and interpret data

❻ Use your results to calculate the number of moles of copper and the number of moles of oxygen (A_r: O = 16; Cu = 64)

❼ What is the ratio of moles of copper: moles of oxygen? (Write as Cu:O 1: ?)

Evaluate data and methods

❽ What do your calculations suggest is the most likely formula of copper(II) oxide?

❾ List the main sources of error in the experiment.

Finding the formula of water

You may have seen an experiment in which an electric current is passed through water (it is actually a dilute solution of sulfuric acid). When the power is turned on, gases from the water collect at the two electrodes. After half an hour about twice as much gas is collected at the negative terminal (**cathode**) as at the positive terminal (**anode**).

The gas collected at the cathode is hydrogen and the gas collected at the anode is oxygen.

As one mole of any gas occupies the same volume at the same temperature and pressure and because twice as much hydrogen is produced as oxygen, this shows that the formula for water must be H_2O (see page 65).

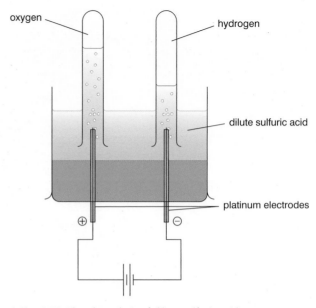

△ Fig. 1.53 The electrolysis of dilute sulfuric acid.

Finding the formula of salts containing water of crystallisation

Some crystals contain molecules of water bound up in the crystal structure. This water is called **water of crystallisation**. Some examples are given below:

Copper(II) sulfate crystals $CuSO_4 \cdot 5H_2O$

Cobalt(II)chloride crystals $CoCl_2 \cdot 6H_2O$

If the crystals are heated, the water of crystallisation can be removed. By weighing the crystals before and after heating, the number of moles of water of crystallisation can be worked out.

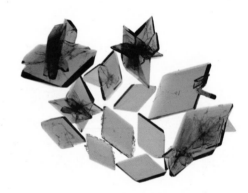

△ Fig. 1.54 Copper(II) sulfate crystals.

WORKED EXAMPLE

2.5 g of blue copper(II) sulfate crystals were heated in a crucible until no further change was observed. The remaining pale blue powder weighed 1.6 g. Calculate the number of moles of water of crystallisation in the crystals. (A_r: H = 1; O = 16; S = 32; Cu = 64)

The copper(II) sulfate crystals were made up of:

	$CuSO_4$	H_2O	
	1.6 g	0.9 g	(Total mass = 2.5 g)
Number of moles	$\dfrac{1.6}{160} = 0.01$	$\dfrac{0.9}{18} = 0.05$	

The ratio of $CuSO_4 : H_2O$ is 1:5. The formula of the crystals is therefore $CuSO_4 \cdot 5H_2O$.

1. These questions refer to the experiment used to find the formula of magnesium oxide:

   a) Why is the lid of the crucible lifted whilst the magnesium is being heated?

   b) Why is the crucible lid lifted only very slightly during the heating of the magnesium?

2. These questions refer to the experiment to find the formula of water:

   a) What material is used for the electrodes?

   b) Which gas forms at the positive electrode (anode)?

USING MOLES OF ATOMS TO FIND CHEMICAL FORMULAE

A chemical formula shows the number of atoms of each element that combine together. For example:

H_2O: A water molecule contains two hydrogen atoms and one oxygen atom.

Alternatively:

H_2O: one mole of water molecules is made from two moles of hydrogen atoms and one mole of oxygen atoms.

The formula of a compound can be calculated if the numbers of moles of the combining elements are known.

WORKED EXAMPLES

1. What is the simplest formula of a hydrocarbon that contains 60 g of carbon combined with 20 g of hydrogen? (A_r: H = 1, C = 12)

	C	H
Write down the mass of each element: (assume 100 g of calcium carbonate)	60	20
Work out the number of moles of each element:	$\frac{60}{12} = 5$	$\frac{20}{1} = 20$
Find the simplest ratio (divide by the smaller number):	$\frac{5}{5} = 1$	$\frac{20}{5} = 4$
Write the formula showing the ratio of atoms:	CH_4	

2. What is the simplest formula of calcium carbonate if it contains 40% calcium, 12% carbon and 48% oxygen? (A_r: C = 12, O = 16, Ca = 40)

	Ca	C	O
Write down the mass of each element:	40	12	48
Work out the number of moles of each element:	$\frac{40}{40} = 1$	$\frac{12}{12} = 1$	$\frac{48}{16} = 3$
Find the simplest ratio:	(Already in the simplest ratio)		
Write the formula showing the ratio of atoms:	$CaCO_3$		

When calculating moles of elements, you must be careful to make sure you know what the question refers to. For example, you may be asked for the mass of 1 mole of nitrogen gas. N = 14 g, but nitrogen gas is diatomic, that is N_2, so the mass of 1 mole N_2 = 28 g. This also applies to other diatomic elements, such as Cl_2, Br_2, I_2, O_2 and H_2.

Formulae such as CH_4 and $CaCO_3$ are called **empirical formulae**. This means that the formula shows the simplest ratio of the atoms present.

Now consider the substance ethane. The formula for ethane is C_2H_6. The simplest ratio of the atoms in ethane would be 1 : 3. But this substance cannot exist. The formula C_2H_6 shows the actual number of atoms of each element in one molecule of ethane.

This is called the **molecular formula**.

The worked examples on pages 60 and 61 show how empirical and molecular formulae can be calculated from experimental data. The questions below provide other examples of the process of taking experimental results and calculating empirical and molecular formulae.

QUESTIONS

1. In an experiment 5.6 g of iron reacts to form 8.0 g of iron oxide. What is the formula of the iron oxide? (A_r: O = 16; Fe = 56)

2. 16.2 g of zinc oxide is reduced by carbon to form 13.0 g of zinc. What is the formula of zinc oxide? (A_r: O = 16; Zn = 65)

3. Butane has an empirical formula of C_2H_5. Its relative formula mass is 58. What is the molecular formula of butane? (A_r: H = 1; C = 12)

4. Hydrogen peroxide has an empirical formula of HO. Its relative formula mass is 34. What is the molecular formula of hydrogen peroxide? (A_r: H= 1; O = 16)

PERCENTAGE YIELD

When you prepare chemicals using chemical reactions, the substances that are formed are called the products. The amount of product you get is called the **yield**.

Sometimes the yield is less than you would expect. There may be some product left behind at different stages in the preparation.

A useful way of comparing the yields from different processes is to find the **percentage yield**. To do this, the chemist measures the actual yield (how much was actually made), works out the predicted yield for the reaction using the balanced symbol equation, and uses this formula:

$$\text{percentage yield} = \frac{\text{actual yield}}{\text{predicted yield}} \times 100$$

WORKED EXAMPLE

If 25 cm^3 of ammonia solution reacts to form the fertiliser ammonium sulfate, the predicted yield is 3.30 g.

However, when chemists actually made this they found 2.64 g was the mass produced.

What is the percentage yield?

$$\text{percentage yield} = \frac{\text{actual yield}}{\text{predicted yield}} \times 100$$

$$= \frac{2.64}{3.30} \times 100$$

$$= 80\%.$$

MOLES OF SOLUTIONS

A solution is made when a **solute** dissolves in a **solvent**. The concentration of a solution depends on how much solute is dissolved in how much solvent. The concentration of a solution is defined in terms of moles per dm^3 (1000 cm^3), or mol/dm^3.

---

1 mole of solute dissolved to make 1000 cm^3 of solution produces a 1 mol/dm^3 solution.

2 moles dissolved to make a 1000 cm^3 solution produces a 2 mol/dm^3 solution.

0.5 moles dissolved to make a 1000 cm^3 solution produces a 0.5 mol/dm^3 solution.

1 mole dissolved to make a 500 cm^3 solution produces a 2 mol/dm^3 solution.

1 mole dissolved to make a 250 cm^3 solution produces a 4 mol/dm^3 solution.

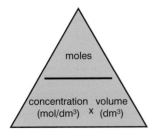

◁ Fig. 1.55 This triangle will help you to calculate concentrations of solutions.

If the same amount of solute is dissolved to make a smaller volume of solution, the solution will be more concentrated.

WORKED EXAMPLES

1. What is the concentration in mol/dm^3 of a solution containing 0.1 moles of copper(II) sulphate in $250cm^3$ of water?

    Step 1: 0.1 moles in $250\,cm^3$

    Step 2: This is equivalent to a solution which contains 0.4 moles in $1000\,cm^3$

    The concentration of the solution is therefore **$0.4\,mol/dm^3$**

2. What is the concentration in mol/dm^3 of a solution containing 1 g of sodium hydroxide (NaOH) in $100\,cm^3$ of water?

    Step 1: 1 mole of NaOH = 23 + 16 + 1 = 40 g

    So 1 g of NaOH = 1/40 = 0.025 mole

    0.025 mole in $100\,cm^3$ of solution

    Step 2: This is equivalent to a solution which contains 0.25 mole in $1000\,cm^3$ of solution.

    The concentration is therefore **$0.25\,mol/dm^3$**

3. Determine the number of moles of hydrochloric acid (HCl) contained in $50\,cm^3$ of a solution with concentration of $0.1\,mol/dm^3$.

    Step 1: $1000\,cm^3$ of $1.0\,mol/dm^3$ solution contains 1 mole of hydrochloric acid

    Step 2: $1000\,cm^3$ of $0.1\,mol/dm^3$ solution therefore contains 0.1 mole of hydrochloric acid

    Step 3: $50\,cm^3$ of $0.1\,mol/dm^3$ solution therefore contains $0.1 \times 50/1000 = 0.005$ mole of hydrochloric acid

    The number of moles of hydrochloric acid = **0.005 moles**.

4. Determine the number of moles of potassium hydroxide (KOH) contained in $200\,cm^3$ of a solution with concentration of $0.5\,mol/dm^3$.

    Step 1: $1000\,cm^3$ of $1.0\,mol/dm^3$ solution contains 1 mole of potassium hydroxide

    Step 2: $1000\,cm^3$ of $0.5\,mol/dm^3$ solution therefore contains 0.5 mole of potassium hydroxide

    Step 3: $200\,cm^3$ of $0.5\,mol/dm^3$ solution therefore contains $0.5 \times 200/1000 = 0.1$ mole of potassium hydroxide

    The number of moles of potassium hydroxide = **0.1 moles**.

5. What mass of sodium chloride can be made from reacting $100\,cm^3$ of $1.0\,mol/dm^3$ hydrochloric acid with excess sodium hydroxide solution? (Na = 23, Cl = 35.5)

Equation:	$HCl(aq)$	+	$NaOH(aq)$	$\rightarrow$	$NaCl(aq)$	+	$H_2O(l)$
Moles:	1		1		1		1
Masses/volumes:	$1000\,cm^3$		$1\,mol/dm^3$		58.5 g		
Therefore	$100\,cm^3$		$1\,mol/dm^3$		5.85 g		

(the quantities are scaled down by 10 times)

QUESTIONS

1. What mass of calcium oxide can be made from the decomposition of 50 g of calcium carbonate? (A_r: C = 12; O = 16; Ca = 40)

Equation: $CaCO_3(s) \rightarrow CaO(s) + CO_2(g)$

2. How many moles of solute are there in the following solutions?

a) 2000 cm³ of 1M sodium chloride solution

b) 100 cm³ of 0.1M copper(II) sulfate solution

c) 500 cm³ of 0.5M sodium hydroxide solution

MOLES OF GASES

In reactions involving gases it is often more convenient to measure the **volume** of a gas rather than its mass.

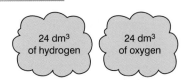

△ Fig. 1.56 Each of these contains 1 mole (6×10^{23}) of molecules.

There are many gases and they are crucially important in science. In experiments or industrial processes it's often necessary to know the amount of a gas, but gas is difficult to weigh. Molar volumes make it possible to find out the amount of a gas by using volume rather than mass.

The volume of one mole of any gas contains the Avogadro constant number of molecules (particles) of that gas. This means that equal volumes of all gases taken at the same temperature and pressure must contain the same number of molecules. This is sometimes called **Avogadro's law**.

One mole of any gas occupies the same volume under the same conditions of temperature and pressure. The conditions chosen are usually room temperature (25 °C) and normal atmospheric pressure (1 atmosphere).

One mole of any gas occupies 24 000 cm³ (24 dm³) at room temperature and pressure (rtp). The following equation can be used to convert gas volumes into moles:

$$moles = \frac{volume\ in\ cm^3}{24\ 000}$$

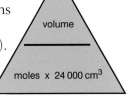

△ Fig. 1.57 The triangle can be used to work out whether to multiply or divide the quantities.

QUESTIONS

1. How many moles of molecules are there in the following:

a) 88 g of carbon dioxide, CO_2? (A_r: C = 12; O = 16)

b) 9 g of water, H_2O? (A_r: H = 1; O = 16)

c) 2.8 g of ethene, C_2H_4? (A_r: H = 1; C = 12)

2. How many moles of molecules are there in the following:

a) 12 000 cm³ of sulfur dioxide, SO_2, at room temperature and pressure?

b) 2400 cm³ of methane, CH_4, at room temperature and pressure?

c) 48 000 cm³ of oxygen, O_2, at room temperature and pressure?

WORKED EXAMPLES

1. What volume of hydrogen is formed at room temperature and pressure when 4 g of magnesium is added to excess dilute hydrochloric acid? (A_r: Mg = 24, molar volume at room temperature and pressure = 24 000 cm^3)

Equation: $Mg(s) + 2HCl(aq) \rightarrow MgCl_2(aq) + H_2(g)$

Moles: 1   2   1   1

Masses/volumes: 24 g   24 000 cm^3

Number of moles magnesium = $\dfrac{4}{24}$ = 0.167

Therefore 0.167 moles hydrogen gas will be produced.

Therefore volume of hydrogen gas = 0.167 × 24 000

$$= 4000 \text{ cm}^3 \text{ at rtp}$$

Note: The hydrochloric acid is in excess. This means that there is enough to react with all the magnesium.

2. What volume of carbon dioxide will be produced when 124 g of copper carbonate is broken down by heating?
(A_r = C = 12, 0 = 16, Cu = 64, molar volume at room temperature and pressure = 24 000 cm^3)

Words: copper carbonate → copper oxide + carbon dioxide

Equation: $CuCO_3(s)$ → $CuO(s)$ + $CO_2(g)$

Moles: 1   1   1

Masses/vols: 124 g   24 000 cm^3

The relative formula mass of $CuCO_3$ is 64 + 12 + 16 + 16+ 16 = 124 g
As there is exactly one mole of reactant, there must be one mole of products produced. One mole of CO_2 is produced or 24 000 cm^3 at rtp.

QUESTIONS

1. Magnesium reacts with steam producing magnesium oxide and hydrogen as shown in the equation:

$Mg(s) + H_2O(g) \rightarrow MgO(s) + H_2(g)$

What volume of hydrogen, measured at room temperature and pressure, would be produced from 4g of magnesium and excess steam? (A_r: Mg = 24; molar volume of a gas = 24 dm^3 at rtp.)

2. Hydrogen peroxide decomposes to form oxygen and water as shown in the equation:

$2H_2O_2(aq) \rightarrow 2H_2O(l) + O_2(g)$

How many moles of hydrogen peroxide would be needed to produce 12 000 cm^3 of oxygen, measured at room temperature and pressure? (Molar volume of a gas = 24 dm^3 at rtp.)

End of topic checklist

The **relative formula mass** is the sum of the relative atomic masses of the constituent atoms.

A **mole** is an amount of a substance containing 6×10^{23} particles (atoms, molecules, ions).

The **empirical formula** of a compound is the simplest formula that shows the whole number ratio of the atoms in the compound.

The **molecular formula** of a compound shows the actual number of atoms in the compound.

The **percentage yield** gives the proportion of the actual amount of product formed compared to the expected amount (as predicted by the equation) in a chemical reaction.

The facts and ideas that you should know and understand by studying this topic:

○ Be able to write word equations and balanced symbol equations, including state symbols, for a range of chemical reactions.

○ Be able to calculate relative formula masses (M_r) from relative atomic masses (A_r).

○ Be able to use the mole to represent amount of substance.

○ Be able to carry out calculations involving amount of substance, relative atomic mass (A_r) and relative formula mass (M_r).

○ Understand how the formulae of simple chemical compounds can be worked out from experimental results for:
  ● metal oxides
  ● water
  ● salts containing water of crystallisation.

○ Be able to calculate empirical and molecular formulae from experimental data.

○ Be able to use chemical equations to calculate reacting masses and the amounts of products formed.

○ Know that the molar gas volume for any gas is 24 dm³ (24 000 cm³) at room temperature and pressure (rtp) and use this in calculations involving gases.

○ Understand how to carry out mole calculations using volumes and molar concentrations (in mol/dm³) of solutions.

○ Be able to calculate percentage yields in reactions by comparing actual amounts produced with those predicted by the chemical equation.

○ Be able to determine the formula of a metal oxide by combustion (e.g. magnesium oxide) or by reduction (e.g. copper(II) oxide).

End of topic questions

1. Write symbol equations from the following word equations:

   a) carbon + oxygen → carbon dioxide   (1 mark)

   b) iron + oxygen → iron(III) oxide (Fe_2O_3)   (1 mark)

   c) iron(III) oxide + carbon → iron + carbon dioxide   (1 mark)

   d) calcium carbonate + hydrochloric acid → calcium chloride + carbon dioxide + water.   (1 mark)

2. What is the formula mass of the following?

   a) ethene, C_2H_4 (1 mark)

   b) sulfur dioxide, SO_2 (1 mark)

   c) methanol, CH_3OH (1 mark)

   (A_r: H = 1; C = 12; O = 16: S = 32)

3. How many moles are in the following?

   a) 64 g of S_8 (1 mark)

   b) 9.8 g of H_2SO_4 (1 mark)

   c) 21 g of Li   (1 mark)

   (A_r: S = 32; H = 1; O = 16; Li = 7)

4. What is the mass of the following?

   a) 2.5 moles of Sr   (1 mark)

   b) 0.25 moles of MgO   (1 mark)

   c) 0.1 moles of C_2H_5Br (1 mark)

   (Sr = 88; Mg = 24; O = 16; C = 12; H = 1; Br = 80)

5. 0.64 g of copper when heated in air form 0.80 g of copper oxide. What is the simplest formula of copper oxide?   (2 marks)

   (A_r: O = 16; Cu = 64)

6. Calculate the simplest formulae of the compounds formed in the following reactions:

   a) 2.3 g of sodium reacting with 8.0 g of bromine.   (2 marks)

   b) 0.6 g of carbon reacting with oxygen to make 2.2 g of compound.   (2 marks)

   c) 11.12 g of iron reacting with chlorine to make 32.20 g of compound. (2 marks)

   (A_r: C = 12; O = 16; Na = 23; Cl = 35.5; Fe = 56; Br = 80)

7. Titanium chloride contains 25% titanium and 75% chlorine by mass. Work out the simplest formula of titanium chloride. (A_r: Ti = 48, Cl = 35.5) **(3 marks)**

8. Ethene has an empirical formula of CH_2 and a relative formula mass of 28. What is the molecular formula of ethene? **(2 marks)**

9. A hydrocarbon contains 92.3% carbon and 7.7% hydrogen.

   a) What is its empirical formula? **(2 marks)**

   b) Its relative formula mass is 26. What is its molecular formula? **(2 marks)**

10. What mass of sodium hydroxide can be made by reacting 2.3 g of sodium with water? **(3 marks)**

    (A_r: H = 1, O = 16, Na = 23)

    $2Na(s) + 2H_2O(l) \rightarrow 2NaOH(aq) + H_2(g)$

11. What mass of barium sulfate can be produced from 50 cm^3 of 0.2 mol/dm^3 barium chloride solution and excess sodium sulfate solution? **(3 marks)**

    (A_r: O = 16, S = 32, Ba = 137)

    $BaCl_2(aq) + Na_2SO_4(aq) \rightarrow BaSO_4(s) + 2NaCl(aq)$

12. In an experiment to make calcium oxide the predicted yield was 2.8 g. The actual yield was 2.1 g. Calculate the percentage yield achieved in the experiment. **(2 marks)**

13. How many moles are in the following?

    a) 24 000 cm^3 of hydrogen gas, measured at room temperature and pressure **(1 mark)**

    b) 1200 cm^3 of nitrogen gas measured at room temperature and pressure **(1 mark)**

14. Iron(III) oxide is reduced to iron by carbon monoxide.

    (A_r: C = 12, O = 16, Fe = 56)

    $Fe_2O_3(s) + 3CO(g) \rightarrow 2Fe(s) + 3CO_2(g)$

    a) Calculate the mass of iron that could be obtained by the reduction of 800 tonnes of iron(III) oxide. **(3 marks)**

    b) What volume of carbon dioxide, measured at room temperature and pressure, would be obtained by the reduction of 320 g of iron(III) oxide? **(3 marks)**

15. Magnesium burns in oxygen to form magnesium oxide:

$$2Mg(s) + O_2(g) \rightarrow 2MgO(s)$$

(A_r: O = 16; Mg = 24)

Calculate:

a) the mass of magnesium required to make 8 g of magnesium oxide **(3 marks)**

b) the mass of oxygen required to make 8 g of magnesium oxide. **(1 mark)**

16. 1 g of magnesium ribbon is added to 100 cm³ of 1M hydrochloric acid. The equation for the reaction is:

$$Mg(s) + 2HCl(aq) \rightarrow MgCl_2(aq) + H_2(g)$$

(A_r: H = 1; Cl = 35.5; Mg = 24. 1 mole of gas occupies 24 000 cm³ at room temperature and pressure.)

a) How many moles of magnesium were used in the experiment? **(1 mark)**

b) How many moles of hydrochloric acid were used in the experiment? **(1 mark)**

c) Which of the reactants, magnesium or hydrochloric acid, was in excess (that is, there was more than needed to complete the reaction and some would be left at the end of the reaction)? Explain how you worked out your answer. **(3 marks)**

d) What mass of magnesium chloride should be produced in the experiment? You must show your working. **(3 marks)**

e) What mass of hydrogen should be produced in the experiment? Show you working. **(3 marks)**

f) What volume of hydrogen, measured at room temperature and pressure, should be made in the reaction? **(2 marks)**

g) After the reaction was completed a student took the solution of magnesium chloride and left it in the sun to crystallise. She then dried the crystals and weighed them. She found that there were 3.1 g of crystals. Use your answer to part **d)** to work out the percentage yield of the magnesium chloride. **(2 marks)**

Ionic bonding

INTRODUCTION

When the atoms of elements react and join together, they form compounds. When one of the reacting atoms is a metal, the compounds formed are called ionic compounds. Ionic compounds do not contain molecules; instead they are made of particles called ions. Ionic compounds all have common properties, many of which are quite different from those of compounds made up of atoms or molecules.

△ Fig. 1.58 Sodium chloride is an example of an ionic compound.

KNOWLEDGE CHECK

✓ Understand that compounds are formed when the atoms of two or more elements combine together.
✓ Know that protons have a positive charge and are found in the nucleus of the atom.
✓ Know that electrons have a negative charge and are found in shells around the nucleus.
✓ Know that the number of outer electrons in an atom depends on its group in the Periodic Table.

LEARNING OBJECTIVES

✓ Understand how ions are formed by atoms gaining or losing electrons.
✓ Know that the ions in an ionic compound are held together by strong electrostatic forces.
✓ Be able to explain why compounds with giant ionic lattices have high melting and boiling points.
✓ Know that ionic compounds do not conduct electricity when solid, but do conduct electricity when molten and in aqueous solution.
✓ Be able to draw 'dot and cross' diagrams showing how ionic compounds are formed by electron transfer in combinations of elements from Groups 1, 2, 3 and 5, 6, 7.
✓ Know the charges of a range of common metal and non-metal ions.
✓ Write formulae for compounds formed between these ions.

THE FORMATION OF IONS

Atoms bond together with other atoms in a chemical reaction to make a compound. For example, sodium reacts with chlorine to make sodium chloride. Ionic compounds contain a metal combined with one or more non-metals. They are not made up of molecules – they are made up of **ions**.

Ions are formed from atoms by the gain or loss of electrons. Both metals and non-metals become more stable by loss or gain of electrons to obtain a noble gas configuration.

Metals lose electrons from their outer shells and form positive ions.

Non-metals gain electrons into their outer shells and form negative ions.

The positive and negative ions are strongly attracted to each other. These attractions are known as electrostatic attractions and are the basis of ionic bonding.

The bonding process can be represented in **dot and cross diagrams**. Look at the reaction between sodium and chlorine as an example.

Sodium is a metal. It has an atomic number of 11 and so has 11 electrons, arranged 2, 8, 1. Its atom diagram looks like this:	Chlorine is a non-metal. It has an atomic number of 17 and so has 17 electrons, arranged 2, 8, 7. Its atom diagram looks like this:

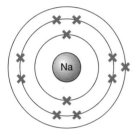 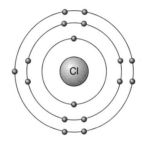

△ Fig. 1.59 Dot and cross diagram for sodium and chlorine.

Sodium has one electron in its outer shell. It can achieve a full outer shell by losing this electron. The sodium atom transfers its outermost electron to the chlorine atom.	Chlorine has seven electrons in its outer shell. It can achieve a full outer shell by gaining an extra electron. The chlorine atom accepts an electron from the sodium.

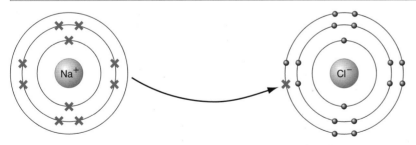

△ Fig. 1.60 Dot and cross diagram for sodium chloride, NaCl.

The sodium is no longer an atom; it is now an ion. It does not have equal numbers of protons and electrons, so it is no longer neutral. It has one more proton than it has electrons, so it is a positive ion with a charge of 1+. The ion is written as Na^+.	The chlorine is no longer an atom; it is now an ion. It does not have equal numbers of protons and electrons, so it is no longer neutral. It has one more electron than protons, so it is a negative ion with a charge of 1−. The ion is written as Cl^-.

METALS CAN TRANSFER MORE THAN ONE ELECTRON TO A NON-METAL

Magnesium combines with oxygen to form magnesium oxide. The magnesium atom (electron arrangement 2, 8, 2) transfers two electrons to the oxygen atom (electron arrangement 2, 6). Magnesium therefore forms a Mg^{2+} ion and oxygen forms an O^{2-} ion.

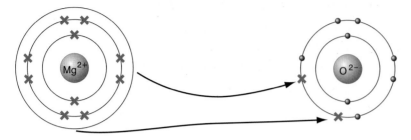

△ Fig 1.61 Dot and cross diagram for magnesium oxide, MgO.

Aluminium has an electron arrangement 2, 8, 3. When it combines with fluorine with an electron arrangement 2, 7, three fluorine atoms are needed for each aluminium atom. The formula of aluminium fluoride is therefore AlF_3.

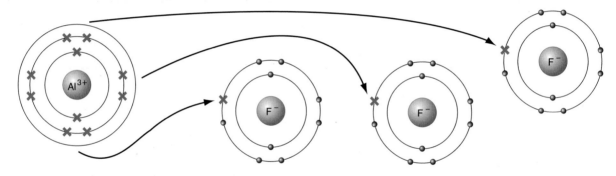

△ Fig. 1.62 Dot and cross diagram for aluminium fluoride, AlF_3.

REMEMBER

In ionic bonding the oxygen atom gains two electrons and changes into the oxide ion (O^{2-}). In this way, the oxygen atom is *reduced* by the addition of the electrons. The magnesium atom loses two electrons, so it is *oxidised*.

SCIENCE IN CONTEXT

MAGNESIUM OXIDE

Magnesium oxide is a very useful and versatile compound. It is used extensively in the construction industry, both in making cement and in making fire-proof construction materials – the fact that it has a melting point of over 2800 °C makes it ideal for this use. It is also used in medicine to relieve sore stomachs and indigestion. Historically it was known as magnesia.

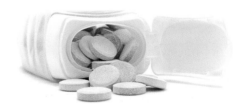

△ Fig. 1.63 These tablets for indigestion contain magnesium oxide.

QUESTIONS

Note: in the examination you need only show the outer electrons in dot and cross diagrams.

1. Draw a dot and cross diagram to show how lithium and fluorine atoms combine to form lithium fluoride. You must show the starting atoms and the finishing ions. (Atomic numbers: Li 3; F 9)

2. Draw a dot and cross diagram to show how calcium and sulfur atoms combine to form calcium sulfide. You must show the starting atoms and finishing ions. (Atomic numbers: Ca 20; S 16)

3. Would you expect the bonding in phosphorus oxide to be ionic? Explain your answer.

IONIC CHARGES

When atoms form ions, they become more stable by achieving the electronic configuration of their nearest noble gas. The charges on some common ions are shown in Table 1.13.

Element or radical	Charge on ion	Example
Group 1 elements	+1	Na^+
Group 2 elements	+2	Mg^{2+}
Group 3 elements	+3	Al^{3+}
Group 5 elements	−3	N^{3-}
Group 6 elements	−2	O^{2-}
Group 7 elements	−1	Cl^-
Transition and other metals	+1 or +2 or +3	Ag^+, Cu^{2+}, Fe^{2+}, Fe^{3+}, Pb^{2+}, Zn^{2+}
Other elements and radicals	+1 or −1 or −2	hydrogen H^+, hydroxide OH^-, ammonium NH_4^+, carbonate CO_3^{2-} nitrate NO_3^-, sulfate SO_4^{2-}

△ Table 1.13 Charges on some common ions and radicals.

These ionic charges can be used to write the chemical formulae of ionic compounds. It is important to remember that the chemical formula has no overall charge and so the charges on the positive ions must cancel the charges on the negative ions.

WORKED EXAMPLES

1. State the formula of calcium chloride given constituent ions as Ca^{2+} and Cl^-.
   Step 1: The positive and negative ion charges do not cancel – two chloride ions are needed to cancel the charge on one calcium ion.
   Step 2: The formula is therefore $CaCl_2$. (Note: no brackets are required as Cl is the symbol for one element, chlorine.)

2. State the formula of iron(III) sulfate given the constituent ions as Fe^{3+} and SO_4^-.

Step 1: The positive and negative ion charges do not cancel out – two iron ions are needed to cancel the charge on three sulfate ions.

Step 2: Brackets will be needed for the sulfate ion as it contains more than one element.

Step 3: The formula is therefore $Fe_2(SO_4)_3$.

PROPERTIES OF IONIC COMPOUNDS

Ionic compounds have high melting points and high boiling points because of strong electrostatic forces between ions.

The strong electrostatic attraction between oppositely charged ions is called an **ionic bond**.

Ionic compounds form **giant ionic lattices**. A lattice is a 3D arrangement of particles (in this case positive and negative ions). In an ionic lattice the positive ions are surrounded by negative ions and the negative ions are surrounded by positive ions. For example, as shown in Fig 1.64, when sodium chloride is formed by ionic bonding, the ions do not pair up. Each sodium ion is surrounded by six chloride ions, and each chloride ion is surrounded by six sodium ions.

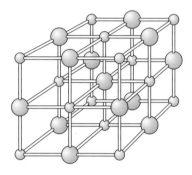

○ chloride ion   ○ sodium ion

△ Fig. 1.64 In solid sodium chloride, the ions are held firmly in place. Ionic compounds have giant ionic lattice structures like this.

The electrostatic attractions between the ions are very strong. The properties of sodium chloride can be explained using this model of its structure.

Properties of sodium chloride	Explanation in terms of structure
Hard crystals	Strong forces of attraction between the ions
High melting point (801°C)	Strong forces of attraction between the ions
Dissolves in water	The water is also able to form strong electrostatic attractions with the ions – the ions are 'plucked' off the lattice structure
Does not conduct electricity when solid	Strong forces between the ions prevent them from moving
Does conduct electricity when molten or dissolved in water	The strong forces between the ions have been broken down and so the ions are able to move

△ Fig. 1.65 Crystals of sodium chloride.

△ Table 1.14 Properties of sodium chloride.

Magnesium oxide is another ionic compound. Its ionic formula is $Mg^{2+}O^{2-}$.
MgO has a much higher melting point and boiling point than NaCl because of the increased charges on the ions. The forces holding the ions together are greater in MgO than in NaCl. Aluminium oxide, Al_2O_3, made up of Al^{3+} ions and O^{2-} ions, has a still higher melting and boiling point due to the higher charge on the aluminium ion.

SCIENCE IN CONTEXT

IONIC CRYSTALS

All ionic compounds form giant structures, and all have relatively high melting and boiling points. The charges on the ions determine the strength of the electrostatic attraction between the ions and so the melting and boiling points of the compound compared to others.

Another factor that affects the strength of the electrostatic attraction is the relative sizes of the positive and negative ions and how well they are able to pack together. The overall arrangement of the ions in the structure is determined by the attractive forces between oppositely charged ions and repulsive forces between similarly charged ions. In sodium chloride, for example, six chloride ions fit around one sodium ion without the chloride ions getting too close together and repelling one another; similarly, six sodium ions can fit around one chloride ion. This structure is sometimes called a 6:6 lattice (see page 76 for a diagram of this structure). Caesium is a metal in the same group of the Periodic Table as sodium, but the caesium ion is much bigger than the sodium ion. In the structure of caesium chloride, eight chloride ions can fit around each caesium ion. So although sodium and caesium are in the same group, their chlorides have very different structures.

△ Fig. 1.66 These gemstones are ionic crystals.

Some of the most valuable gemstones are ionic compounds. Rubies and sapphires, for example, are both aluminium oxide. The different colours of the gemstones are due to traces of other metals such as iron, titanium and chromium.

QUESTIONS

1. Why does an ionic compound such as magnesium oxide not conduct electricity when it is solid?

2. Suggest a reason why magnesium oxide has a higher melting point than sodium chloride.

End of topic checklist

An **ionic** compound is formed by the reaction between a metal and one or more non-metals.

In the solid state the ions in an ionic compound are arranged in a giant three-dimensional lattice structure held together by the attractions between oppositely charged ions.

The facts and ideas that you should know and understand by studying this topic:

○ Be able to describe the formation of ions by the gain or loss of electrons.

○ Know the charges of these ions:

- metals in Groups 1, 2 and 3
- non-metals in Groups 5, 6 and 7
- Ag^+, Cu^{2+}, Fe^{2+}, Fe^{3+}, Pb^{2+}, Zn^{2+}
- hydrogen (H^+), hydroxide (OH^-), ammonium (NH_4^+)
- carbonate (CO_3^{2-}), nitrate (NO_3^-), sulfate (SO_4^{2-})

○ Write formulae for compounds formed between the ions listed above.

○ Be able to explain the formation of ionic compounds of metals from Groups 1, 2, 3 and non-metals from Groups 5, 6, 7 by using dot and cross diagrams.

○ Understand that ionic bonding is a strong electrostatic attraction between oppositely charged ions.

○ Understand that ionic compounds have high melting and boiling points because of strong electrostatic forces between oppositely charged ions.

○ Know that ionic compounds do not conduct electricity when solid, but do conduct when molten and in aqueous solution.

End of topic questions

1. For each of the following reactions, say whether the compound formed is ionic or not:

   a) hydrogen and chlorine   (1 mark)

   b) carbon and hydrogen   (1 mark)

   c) sodium and oxygen   (1 mark)

   d) chlorine and oxygen   (1 mark)

   e) calcium and bromine.   (1 mark)

2. Write down the formulae of the ions formed by the following elements:

   a) potassium   (1 mark)

   b) aluminium   (1 mark)

   c) sulfur   (1 mark)

   d) fluorine.   (1 mark)

3. What are the charges on the ions of the following elements or radicals:

   a) silver   (1 mark)

   b) zinc   (1 mark)

   c) hydroxide   (1 mark)

   d) nitrate.   (1 mark)

4. Write the chemical formulae of the following ionic compounds:

   a) silver nitrate   (1 mark)

   b) zinc hydroxide   (1 mark)

   c) zinc nitrate.   (1 mark)

5. Draw dot and cross diagrams to show how the following atoms combine to form ionic compounds. (You must show the electronic arrangements of the outer shells for the starting atoms and the finishing ions.)

   a) potassium and oxygen (Atomic numbers: K 19; oxygen 8)   (2 marks)

   b) magnesium and chlorine (Atomic numbers: Mg 12; Cl 17)   (2 marks)

6. Explain why an ionic substance such as potassium chloride:

   a) has a high melting point   (2 marks)

   b) behaves as an electrolyte.   (2 marks)

7. Under what conditions do ionic compounds conduct electricity?   (2 marks)

△ Fig. 1.67 All plastics are covalent substances.

Covalent bonding

INTRODUCTION

Unlike ionic compounds, covalent substances are formed when atoms of non-metals combine. Although covalent substances all contain the same type of bonds, their properties can be quite different – some are gases, others are very hard solids with high melting points. Plastics are a common type of covalent substance. Because chemists now understand how the molecules form and link together, they can produce plastics with almost the perfect properties for a particular use, from soft and flexible (as in contact lenses) to hard and rigid (as in electrical sockets).

KNOWLEDGE CHECK

✓ Understand that molecules are made up of two or more atoms combined together.
✓ Know what the atomic number is.
✓ Know that electrons are found in shells around the nucleus of an atom.
✓ Know that the number of electrons in the outer shell of an atom is governed by their group in the Periodic Table.

LEARNING OBJECTIVES

✓ Understand that a covalent bond is formed when a pair of electrons is shared between atoms.
✓ Know that the atoms in a covalent bond are held together by strong electrostatic forces.
✓ Be able to show the formation of simple covalent compounds using dot and cross diagrams, including:
  • hydrogen, oxygen, nitrogen, halogens and hydrogen halides
  • water, ammonia and carbon dioxide
  • methane, ethane, ethene and organic molecules containing halogen atoms.
✓ Be able to explain why substances with a simple molecular structures are gases or liquids, or solids with low melting and boiling points.
✓ Be able to explain why the melting and boiling points of substances with simple molecular structures increase, in general, with increasing relative molecular mass.
✓ Be able to explain why substances with giant covalent structures are solids with high melting and boiling points.
✓ Be able to explain how the structures of diamond, graphite and C_{60} fullerene influence their physical properties, including electrical conductivity and hardness.
✓ Know that covalent compounds do not usually conduct electricity.

HOW COVALENT BONDS ARE FORMED

Covalent bonding involves electron sharing and occurs between atoms of non-metals. It results in the formation of a molecule. The non-metal atoms try to achieve complete outer electron shells by sharing electrons.

A single **covalent bond** is formed when two atoms each contribute one electron to a shared pair of electrons. For example, hydrogen gas exists as H_2 molecules. Hydrogen gas is **diatomic**, that is, it contains two atoms per molecule. Each hydrogen atom needs to fill its electron shell. They can do this by sharing electrons.

represented as

H—H

△ Fig. 1.68 The dot and cross diagram and displayed formula of H_2.

A covalent bond is the result of electrostatic attraction between the bond pair of electrons (negative charges) and the nuclei (positive charges) of the atoms involved in the bond. A single covalent bond can be represented by a single line. The formula of the molecule can be written as a displayed formula, H—H. The hydrogen and oxygen atoms in water are also held together by single covalent bonds.

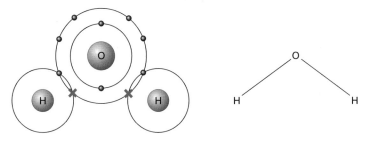

△ Fig. 1.69 Water contains single covalent bonds.

The hydrogen and carbon atoms in methane are held together by single covalent bonds.

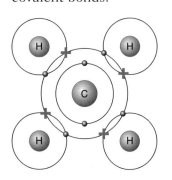

△ Fig. 1.70 Methane contains four single covalent bonds.

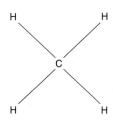

△ Fig. 1.71 The displayed formula for methane.

The hydrogen chloride molecule, HCl, is also held together by a single covalent bond. Hydrogen chloride is an example of a hydrogen halide.

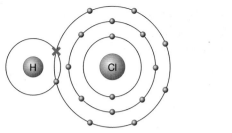

H ———————— Cl

△ Fig. 1.72 Hydrogen chloride has a single covalent bond

Ethane has a slightly more complex electron arrangement.

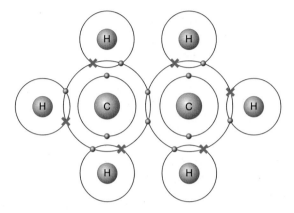

△ Fig. 1.73 Covalent bonds in ethane.

Some molecules contain double covalent bonds. In carbon dioxide, the carbon atom has an electron arrangement of 2, 4 and needs an additional four electrons to complete its outer electron shell. It needs to share its four electrons with four electrons from oxygen atoms (electron arrangement 2, 6). So two oxygen atoms are needed, each sharing two electrons with the carbon atom.

H   H
| |
H—C—C—H
| |
H   H

△ Fig. 1.74 Displayed formula for ethane.

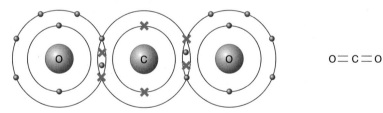

△ Fig. 1.75 Carbon dioxide contains double bonds.

Some molecules contain triple covalent bonds. In the nitrogen molecule, each nitrogen atom has an electron arrangement of 2, 5 and needs an additional three electrons to complete its outer electron shell. It needs to share three of its outer electrons with another nitrogen atom. This forms a triple bond which is shown as N ≡ N.

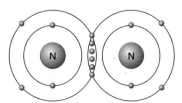

△ Fig. 1.76 A nitrogen molecule contains a triple bond.

It is important to be able to show dot and cross diagrams for halogens such as chlorine, for oxygen, for ammonia and for the hydrocarbon ethene. These diagrams will be produced in answer to questions 1–4 below.

QUESTIONS

1. Draw a dot cross diagram and displayed formula to show how the covalent bonds are formed in chlorine gas (Cl_2). The atomic number of chlorine is 17.

2. Draw a dot and cross diagram and displayed formula to show how the covalent bonds are formed in the gas ammonia (NH_3). The atomic number of hydrogen is 1; the atomic number of nitrogen is 7.

3. Draw a dot and cross diagram and displayed formula to show the double bond in oxygen gas (O_2). The atomic number of oxygen is 8.

4. Draw a dot and cross diagram and displayed formula to show the covalent bonds in ethene (C_2H_4). The atomic number of hydrogen = 1; the atomic number of carbon = 6.

5. Draw a dot and cross diagram and displayed formula to show the covalent bonds in hydrazine (N_2H_4). The atomic number of hydrogen = 1; the atomic number of nitrogen = 7.

HOW MANY COVALENT BONDS CAN AN ELEMENT FORM?

The number of covalent bonds a non-metal atom can form is linked to its position in the Periodic Table. Metals (Groups 1, 2, 3) do not form covalent bonds. The noble gases in Group 0, for example helium, neon and argon, are unreactive and do not usually form covalent bonds either.

Group in the Periodic Table	1	2	3	4	5	6	7	0
Covalent bonds formed	X	X	X	4	3	2	1	X

△ Table 1.15 Group number and number of covalent bonds formed.

MOLECULAR CRYSTALS

Covalent compounds can form simple molecular crystals. Many covalent crystals only exist in the solid form at low temperatures. Some simple molecular crystals are ice, solid carbon dioxide and iodine.

PROPERTIES OF COVALENT COMPOUNDS

Substances with molecular structures are usually gases, liquids or solids with low melting points and boiling points.

Covalent bonds are strong bonds. They are **intramolecular** bonds – formed *within* each molecule. Much weaker **intermolecular** forces attract the individual molecules to each other.

The properties of covalent compounds can be explained using a simple model involving these two types of bond or forces.

Properties of hydrogen	Explanation in terms of structure
Hydrogen is a gas with a very low melting point ($-259\,°C$).	The intermolecular forces of attraction between the molecules are weak.
Hydrogen does not conduct electricity.	There are no ions or free electrons present. The covalent bond (intramolecular bond) is a strong bond and the electrons cannot be easily removed from it.

△ Table 1.16 Properties of hydrogen.

As the relative molecular mass of simple molecular structures increases, in general the intermolecular forces will also increase and so too will the melting and boiling points. Specific examples of this will be seen in the Organic chemistry section on alkanes and alkenes.

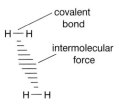

△ Fig 1.77 Intermolecular force between hydrogen molecules.

QUESTIONS

1. Why does a covalently bonded compound such as carbon dioxide have a relatively low melting point?

2. Would you expect a covalently bonded compound such as ethanol to conduct electricity? Explain your answer.

DIAMOND, GRAPHITE AND FULLERENES

Some covalently bonded compounds do not exist as simple molecular structures in the way that hydrogen does. Diamond, for example, exists as a giant structure with each carbon atom covalently bonded to four others. Other forms of carbon include graphite and a group of chemicals called **fullerenes**. One particular fullerene is known as Buckminsterfullerene (named after an architect who first designed geodesic dome buildings with the fullerene structure) and has the formula C_{60}. The structures of these three forms of carbon are shown in Fig. 1.79, 1.81 and 1.82. Different forms of the same element, like these, are called **allotropes**. Carbon has an electron configuration of 2,4 and so has four bonding electrons and can form four covalent bonds. In diamond, each carbon atom forms four strong covalent bonds. In graphite, each carbon atom forms three strong covalent bonds. The fourth bonding electron does not form a covalent bond but instead forms a delocalised cloud of electrons with other non-bonding carbon electrons. There are weak forces of attraction between the layers. The molecules of Buckminsterfullerene are spherical and are also known as 'buckyballs'.

△ Fig. 1.78 A cut diamond.

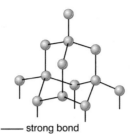
—— strong bond
△ Fig. 1.79 Structure of diamond.

△ Fig. 1.80 A piece of graphite.

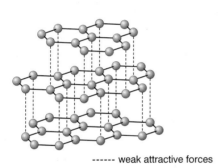
------ weak attractive forces
△ Fig. 1.81 Structure of graphite.

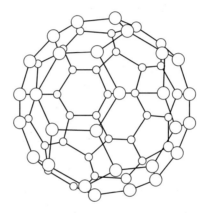
△ Fig. 1.82 The structure of Buckminsterfullerene, C_{60}, is spherical and based on hexagonal rings of carbon atoms.

In diamond, all the bonding is extremely strong, which makes diamond an extremely hard substance – one of the hardest natural substances known. This is why diamonds are used in cutting.

In graphite, carbon atoms form layers of hexagons in the plane of their strong covalent bonds. The weak forces of attraction are between the layers. Because the layers can slide over each other, graphite is flaky and can be used as a lubricant. Graphite can also conduct electricity, because the fourth unbonded electron from each carbon atom is delocalised and so can move along the layer.

The structure of the C_{60} molecule is like a large ball of atoms. Each atom is strongly bonded to three other atoms but, similar to graphite, the fourth electron from each carbon atom is not bonded, it is delocalised. As a result of its structure Buckminsterfullerene is very hard and is a good conductor of electricity.

As the atoms in diamond and graphite are held together by strong covalent bonds they have very high melting points. Diamond, for example, has a melting point of about 3730 °C.

DIAMONDS

Diamond is the hardest naturally occurring material in the world. It is formed by high pressures and high temperatures deep underground. Volcanic eruptions often bring the diamonds closer to the surface, where they can be mined. In some cases the diamonds can be mined almost on the surface of the land, whereas in other cases tunnels need to be dug deep into the ground. Diamonds are mined throughout Africa. The ore is called kimberlite, and diamonds are found in kimberlite gravels and pipe formations.

After mining the ore is crushed, washed and screened by X-ray to find the diamonds. Finally the diamonds are sorted by hand, then washed and classified for sale. Diamond mining and recovery is a very clean operation. Processing of the ore uses no toxic chemicals and produces no chemical pollutants. However, getting the diamonds from deep in the ground can be very dangerous for the miners.

△ Fig. 1.83 This saw has diamonds on its cutting edges.

The quality of diamonds used in jewellery is judged in terms of the four Cs: carat mass, colour, clarity (how transparent the diamond is and how well it reflects light) and cut (the shape of the diamond). One carat is 0.2 g. Although the carat weight is the most important, prices can vary widely depending on the other three factors. In 2009 the largest producers were Russia and Botswana, each producing about 32 million carats of uncut diamonds. In the same year the world demand for diamonds was estimated at 39 billion dollars. Diamonds are clearly big business!

△ Fig. 1.84 An open cast diamond mine at Yakutia in Russia.

QUESTIONS

1. Why does diamond have a very high melting point?

2. How is the structure of graphite different to that of diamond?

End of topic checklist

A **covalent bond** is formed when electrons are shared between the atoms of two non-metals.

Substances with **simple molecular structures** have relatively weak forces of attraction between the molecules and so have low melting and boiling points.

Giant covalent structures contain a lattice of strong covalent bonds.

Fullerenes are **allotropes** (different physical forms) of carbon, the most common being Buckminsterfullerene, C_{60}.

The facts and ideas that you should know and understand by studying this topic:

○ Be able to describe the formation of a covalent bond by the sharing of a pair of electrons between two atoms.

○ Understand that covalent bonding is a result of a strong attraction between the bonding pair of electrons and the nuclei of the atoms involved in the bond.

○ Be able to use dot and cross diagrams for the following:
- diatomic molecules, including hydrogen, oxygen, nitrogen, halogens and hydrogen halides
- inorganic molecules including water, ammonia and carbon dioxide
- organic molecules containing up to two carbon atoms, including methane, ethane, ethene and those containing halogen atoms.

○ Be able to explain why substances with simple molecular structures are gases or liquids, or solids with low melting and boiling points, in terms of the relatively weak forces between the molecules.

○ Be able to explain why the melting and boiling points of substances with simple molecular structures increase, in general, with increasing relative molecular mass.

○ Be able to explain the high melting and boiling points of substances with giant covalent structures in terms of the breaking of many strong covalent bonds.

○ Know that covalent compounds do not usually conduct electricity.

○ Be able to explain how the structures of diamond, graphite and C_{60} fullerene influence their physical properties, including conductivity and hardness.

End of topic questions

1. Draw dot and cross diagrams to show the bonding in the following compounds:

   a) hydrogen fluoride, HF. **(2 marks)**

   b) carbon disulfide, CS_2. **(2 marks)**

   c) ethanol, C_2H_5OH. **(2 marks)**

2. Candle wax is a covalently bonded compound. Explain why candle wax has a relatively low melting point. **(2 marks)**

3. Ozone (O_3) is a gas found in the Earth's atmosphere. How do you know that ozone is covalently bonded and not ionically bonded? **(2 marks)**

4. Explain why methane (CH_4), which has strong covalent bonds between the carbon atom and the hydrogen atoms, is a gas at room temperature and pressure and has a very low melting point. **(2 marks)**

5. Substance X has a simple molecular structure.

   a) In which state(s) of matter might you expect it to exist in at room temperature and pressure? Explain your answer. **(2 marks)**

   b) What sort of boiling point would you expect it to have? Explain your answer. **(2 marks)**

6. Use the structure of graphite to explain:

   a) how carbon fibres can add strength to tennis racquets **(2 marks)**

   b) how graphite conducts electricity. **(2 marks)**

7. a) Diamond is probably the hardest naturally occurring material in the world. Explain this by referring to the structure of diamond. **(2 marks)**

   b) Apart from jewellery, name another use of diamond. **(1 mark)**

8. Fullerenes have exceptional lightness and strength (they are used in the manufacture of tennis racket frames). They are also used as semiconductors in electronic circuits. Explain the properties of strength and electrical conductivity of the fullerenes in terms of their structures. **(2 marks)**

9. Hydrogen chloride and sodium chloride are compounds with different types of chemical bonding. For each compound explain how the atoms are bonded and how the type of bonding causes the differences in the physical properties of the two compounds. **(6 marks)**

Metallic bonding

INTRODUCTION

The structure of a metal does contain an orderly arrangement of atoms and can form crystals. The bonding in metals is different from that in ionic or covalent substances. As with ionic and covalent substances, it is the nature of the bonding in metals that gives them their characteristic properties.

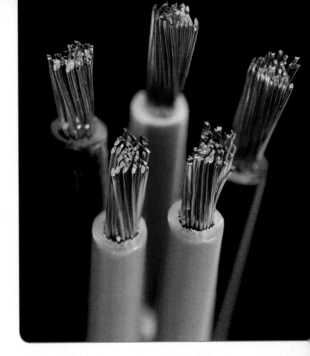

Δ Fig. 1.85 Copper atoms are held together by strong metallic bonds and form a giant structure.

KNOWLEDGE CHECK

✓ Understand the structure and arrangement of protons, neutrons and electrons in an atom.
✓ Be familiar with some of the characteristic properties of metals.

LEARNING OBJECTIVES

✓ Know how to represent a metallic lattice by a 2D diagram.
✓ Understand metallic bonding in terms of electrostatic attractions.
✓ Explain typical physical properties of metals, including electrical conductivity and malleability.

METALLIC BONDING

Metals are giant structures with high melting and boiling points.

Metal atoms give up one or more of their electrons to form positive ions, called **cations**. The electrons they give up form a 'sea of electrons' surrounding the positive metal ions, and the negative electrons are attracted to the positive ions, holding the structure together.

The electrons are free to move through the whole structure. The electrons are **delocalised**, meaning they are not fixed in one position.

THE PROPERTIES OF METALS

Metals are shiny, **malleable** (can be hammered into a sheet), **ductile** (can be drawn or pulled into a wire), good conductors of electricity and good conductors of heat.

Metals are good conductors because their electrons are free to move through the structure. When a metal is in an electric circuit, electrons can move toward the positive terminal and the negative terminal can supply electrons into the metal.

Metals are malleable and ductile because metallic bonds are not as rigid as the bonds in diamond, although they are still very strong. So the ions in the metal can move around into different positions when the metal is hammered or worked.

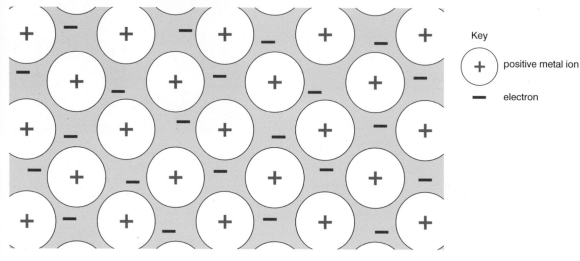

Key

(+) positive metal ion

− electron

△ Fig. 1.86 Positive metal ions and electrons in a metal.

QUESTIONS

1. Why are metals good conductors of electricity?

2. What happens to the structure of a metal when you bend it slightly?

SCIENCE IN CONTEXT FACTS ABOUT METALS

1. The use of metals can be traced back to about 7000 years ago. An archaeological site in Serbia has evidence of the extraction of copper about that time, and gold artefacts dating to about 1000 years later have been found at a burial site at Varna in Bulgaria. Over 2700 years ago seven metals were known and used. These so-called 'metals of antiquity' were gold, copper, silver, lead, tin, iron and mercury. Of these, gold, silver, copper, iron and mercury were found in their native state, that is, as pure elements. (Iron as a pure metal is found only in meteors.)

2. Mercury is the only liquid metal at normal room temperature and pressure.

3. The most reactive metals are found in Group 1 of the Periodic Table and include sodium and potassium.

4. Many of the metals used in construction are found in the middle of the Periodic Table and are called the transition metals.

Δ Fig. 1.87 The Burj Khalifa building in Dubai.

5. Most of the metallic structures around us are not made from pure metals but from **alloys**. Alloys are mixtures of metals or occasionally mixtures of metals and non-metals: for example, steel is an alloy of iron and carbon. Thirty-nine thousand tonnes of steel were used in building The Burj Khalifa building in Dubai (the world's tallest sky scraper at 828 metres tall!) Alloys allow the property of a metal to be modified for a particular purpose. For instance, aluminium is useful in building aircraft because it has a low density, but alloying it with other metals increases its strength. Bronze, an alloy of copper and tin, was the first alloy invented.

6. Some elements have the properties of both metals and non-metals. They are called metalloids. One of the most common metalloids is silicon.

End of topic checklist

Metallic bonding consists of a large 3D array or lattice consisting of positive metal ions and electrons that are free to move held together by electrostatic attractions.

The **malleability** of a substance is a measure of how easily it can be beaten into sheets.

The facts and ideas that you should know and understand by studying this topic:

○ Be able to represent a metallic lattice by a 2D diagram.

○ Understand that a metal can be described as a giant structure of positive ions surrounded by a 'sea' of delocalised electrons.

○ Be able to explain the electrical conductivity of metals in terms of the sea of delocalised electrons.

○ Be able to explain the malleability of a metal in terms of its structure and bonding.

End of topic questions

1. Describe the structure of a metal. (2 marks)

2. Use your knowledge of the structure of a metal to explain why metals:

    a) conduct electricity (2 marks)

    b) can be beaten into sheets (that is, they are malleable). (2 marks)

3. Graphite can conduct electricity in only one plane, but metals can conduct in all planes. Explain why this is so. (3 marks)

4. Use the information in the table to answer the questions that follow.

Substance	Melting point (°C)	Boiling point (°C)	Conductivity	
			Solid	**Molten**
A	751	1244	poor	good
B	−50	148	poor	poor
C	630	1330	good	good
D	1600	2230	poor	poor

Which substance:

    a) is a metal (1 mark)

    b) contains ionic bonds (1 mark)

    c) has a giant covalent structure (1 mark)

    d) has a simple molecular structure? (1 mark)

5. Metals are malleable. Explain what this statement means. (1 mark)

Electrolysis

INTRODUCTION

Most elements in nature are found combined with other elements as compounds. These compounds must be broken down to obtain the elements that they contain. For some, one of the most efficient and economical ways to do this is using electricity in a process called **electrolysis**. Simple electrolysis experiments can be performed in the laboratory, and electrolysis is also used in large-scale industrial processes to produce important chemicals like aluminium and chlorine.

△ Fig. 1.88 Industrial electroplating.

This topic deals with the underlying principles of electrolysis and concentrates on some experiments that can be performed in the laboratory.

KNOWLEDGE CHECK

✓ Know the different arrangements of the particles in solids, liquids and gases.
✓ Understand the terms conductor and insulator.
✓ Understand the differences between ionic and covalent bonding.

LEARNING OBJECTIVES

✓ Be able to explain why covalent compounds do not conduct electricity.
✓ Be able to explain why ionic compounds conduct electricity only when molten or in aqueous solution.
✓ Know that an anion is a negative ion and a cation is a positive ion.
✓ Be able to describe experiments using inert electrodes to investigate electrolysis, and to predict the products.
✓ Know how to write half-equations for the reactions taking place at the electrodes and understand why these reactions are classified as oxidation or reduction.

ELECTROLYTES AND NON-ELECTROLYTES

Compounds that can conduct electricity are called **electrolytes**. Experiments can be carried out using a simple electrical cell, as shown in Fig. 1.89.

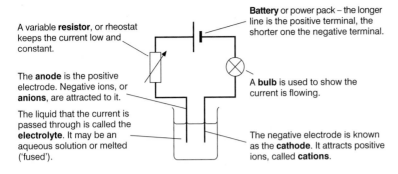

A variable **resistor**, or rheostat keeps the current low and constant.

Battery or power pack – the longer line is the positive terminal, the shorter one the negative terminal.

The **anode** is the positive electrode. Negative ions, or **anions**, are attracted to it.

A **bulb** is used to show the current is flowing.

The liquid that the current is passed through is called the **electrolyte**. It may be an aqueous solution or melted ('fused').

The negative electrode is known as the **cathode**. It attracts positive ions, called **cations**.

△ Fig. 1.89 Simple electrical cell.

When the solution in the beaker is an electrolyte, a complete circuit will form and the bulb will light. The electric current that flows is caused by electrons moving in the electrodes and wires of the circuit and by ions moving in the solution. If the current does not flow, then the beaker must contain a non-electrolyte. Because of this, a simple circuit like this one can be used to distinguish between electrolytes and non-electrolytes.

CONDITIONS FOR ELECTROLYSIS

The substance being electrolysed (the electrolyte) must contain ions, and these ions must be free to move. In other words, the substance must either be molten or dissolved in water.

A d.c. voltage must be used. The **electrode** connected to the positive terminal of the power supply is known as the anode. The electrode connected to the negative terminal of the power supply is known as the cathode. The electrical circuit can be drawn as shown below.

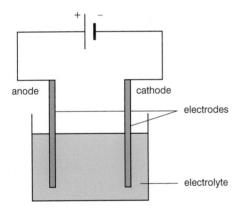

△ Fig. 1.90 A typical electrical circuit used in electrolysis.

Covalent compounds generally do not conduct electricity as they do not contain ions.

However, some covalent compounds, such as hydrogen chloride, form ions when in contact with water. In this example the hydrogen chloride forms hydrochloric acid when in water and this contains H^+ and Cl^- ions.

It is important to remember that in an electrolyte cations are positive ions (attracted to the cathode) and anions are negative ions (attracted to the anode).

QUESTIONS

1. What does the process of electrolysis involve?

2. What is the name given to the positive electrode?

3. What two conditions must apply for a substance to be an electrolyte and allow an electric current to pass through it?

When an electric current passes through an electrolyte, new substances are formed. The examples below show how you can work out what products will form.

ELECTROLYSIS OF MOLTEN LEAD(II) BROMIDE

Lead(II) bromide ($PbBr_2$) is ionically bonded and contains Pb^{2+} ions and Br^- ions. When the solid is melted and a voltage is applied, the ions are able to move. The positive lead ions move to the negative electrode (the cathode), and the negative bromide ions move to the positive electrode (the anode). The electrodes are usually made of carbon, which is inert. This means they do not undergo any chemical change during the electrolysis.

The equations below are ionic **half-equations**. They represent what happens to the ions when they are discharged at each electrode.

At the cathode:

$$Pb^{2+}(l) + 2e^- \rightarrow Pb(l)$$ The lead ions accept electrons to form lead atoms. Accepting electrons is reduction.

At the anode:

$$2Br^-(l) \rightarrow Br_2(g) + 2e^-$$ The bromide ions give up electrons to form bromine molecules. Losing electrons is oxidation.

The products of the electrolysis are lead and bromine. Silvery deposits of lead form near the bottom of the dish, and brown bromine vapour near the anode. In the half-equations above, you will see that the number of electrons accepted and released is the same. The electrical current is produced by this flow of electrons around the external circuit.

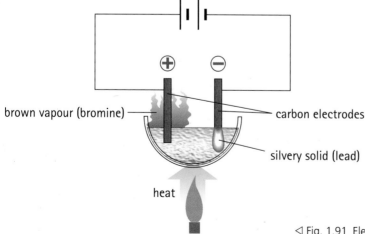

brown vapour (bromine) — carbon electrodes

silvery solid (lead)

heat

◁ Fig. 1.91 Electrolysis of molten lead(II)bromide.

Predicting the products of the electrolysis of simple molten compounds is relatively straightforward. The metal forms at the cathode and the non-metal forms at the anode. For example, the electrolysis of molten aluminium oxide forms aluminium (at the cathode) and oxygen (at the anode).

ELECTROLYSIS OF SODIUM CHLORIDE SOLUTION

This experiment can be performed using a **cell** as shown in the diagram (see Fig. 1.92). Again, inert carbon electrodes are used.

When sodium chloride dissolves in water, the sodium and chloride ions separate and are free to move independently. In addition, the water provides a small quantity of hydrogen (H^+) and hydroxide (OH^-) ions.

$NaCl(aq) \rightarrow Na^+(aq) + Cl^-(aq)$

$H_2O(l) \rightleftharpoons H^+(aq) + OH^-(aq)$

The breaking down of water into ions (also called **dissociation** of water) is a reversible reaction. Although there are very few ions present, if they are removed they will be immediately replaced.

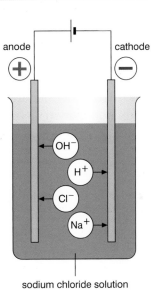

▷ Fig. 1.92 Electrolysis of sodium chloride solution.

Whenever you consider the electrolysis of an aqueous solution, you must always include the H^+ and OH^- ions.

At the cathode   Two ions, Na^+ and H^+, move to the cathode but only H^+ ions are discharged. The sodium ions remain as ions, but the solution turns alkaline as the loss of hydrogen ions leaves a surplus of hydroxide ions.

$2H^+(aq) + 2e^- \rightarrow H_2(g)$ The hydrogen ions accept electrons and form hydrogen molecules. The hydrogen ions have been reduced.

In the case of aqueous solutions there is potential for either the metal or hydrogen to form at the cathode. The product can be determined by consideration of the reactivity series of metals as shown in the adjacent Fig. 1.93. Only metals below hydrogen in the reactivity series are deposited from aqueous solutions; if the metal is more reactive than hydrogen then hydrogen will be produced and not the metal. For example, if aqueous zinc chloride is electrolysed hydrogen will form at the cathode.

At the anode   Two ions, Cl^- and OH^- move to the anode. Both ions could be discharged depending on the concentration of the solution. If the solution is very dilute, OH^- ions are discharged; whereas if the solution is concentrated, Cl^- ions are discharged. At intermediate concentrations both ions are likely to be discharged, giving a mixture of products.

$4OH^-(aq) \rightarrow 2H_2O(l) + O_2(g) + 4e^-$ The hydroxide ions give up electrons and form oxygen molecules.

$2Cl^-(aq) \rightarrow Cl_2(g) + 2e^-$ The chloride ions give up electrons and form chlorine molecules. The chloride ions have been oxidised.

Bubbling or effervescence is seen at each of the two electrodes, and the products of the electrolysis are hydrogen and oxygen and/or chlorine.

Note: The solution becomes alkaline, because the loss of hydrogen ions leaves an excess of hydroxide ions.

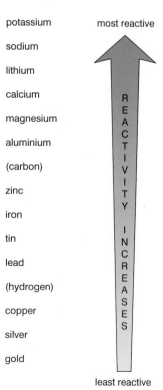

| most reactive |
| potassium |
| sodium |
| lithium |
| calcium |
| magnesium |
| aluminium |
| (carbon) |
| zinc |
| iron |
| tin |
| lead |
| (hydrogen) |
| copper |
| silver |
| gold |
| least reactive |

REACTIVITY INCREASES

Δ Fig. 1.93 Reactivity of metals.

ELECTROLYSIS OF COPPER(II) SULFATE SOLUTION

When copper(II) sulfate is electrolysed using inert platinum electrodes, the following changes occur:

At the cathode	Two ions, Cu^{2+} and H^+, move to the cathode and Cu^{2+} ions are discharged.
$Cu^{2+}(aq) + 2e^- \rightarrow Cu(s)$	The copper ions accept electrons and form copper atoms.
At the anode	Two ions, SO_4^{2-} and OH^-, move to the anode and OH^- ions are discharged.
$4OH^-(aq) \rightarrow 2H_2O(l) + O_2(g) + 4e^-$	The hydroxide ions give up electrons and form oxygen molecules.

The products of the electrolysis are copper and oxygen. The copper forms as a coating on the cathode and bubbles of oxygen are seen next to the anode.

There are potentially a number of different products when aqueous solutions are electrolysed. At the cathode, the product is either the metal or hydrogen. From the reactivity series of metals shown the rule is: only the metals below hydrogen are discharged instead of hydrogen.

For example, if zinc chloride solution is electrolysed, hydrogen, not zinc, is formed at the cathode.

At the anode, the main product often depends on the concentration of the solution. For example, if concentrated hydrochloric acid is electrolysed, chlorine is the main product at the anode. If dilute hydrochloric acid is electrolysed instead, oxygen is likely to be the main product.

REMEMBER

- In electrolysis, negative ions give up electrons and usually form molecules (such as Cl_2, Br_2). Positive ions accept electrons and usually form metallic atoms (such as Cu, Al) or hydrogen gas.

- The loss of electrons is **oxidation** (of the non-metal ions), the gain of electrons is **reduction** (of the metal ions).

QUESTIONS

1. a) What is an inert electrode?

   b) Give an example of a substance that is often used as an inert electrode.

2. What products are formed when the following molten solids are electrolysed?

   a) lead(II) chloride

   b) magnesium oxide

   c) aluminium oxide

3. What are the products at the cathode when the following solutions are electrolysed?

   a) sodium bromide solution

   b) zinc chloride solution

   c) silver nitrate solution

4. a) Write a half-equation showing how oxide ions (O^{2-}) are discharged as oxygen gas.

   b) Which electrode would this change take place at?

Developing investigative skills

The diagram shows the apparatus that can be used to electrolyse dilute sulfuric acid. (Note: dilute sulfuric acid is an irritant to eyes and skin.)

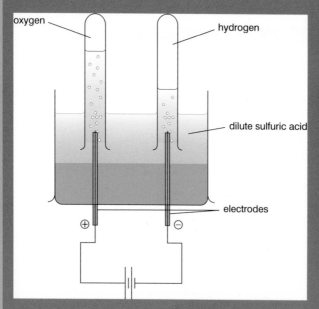

△ Fig. 1.94 Electrolysis of dilute sulfuric acid.

Demonstrate and describe techniques

❶ Inert electrodes were used. Suggest what the electrodes were made of.

❷ Was the hydrogen formed at the anode or the cathode?

❸ Describe how you would put the tubes over the electrodes without losing any of the dilute sulfuric acid in the tubes.

❹ What name is given to solutions such as sulfuric acid that allow an electric current to flow through them?

Make observations and measurements

❺ Oxygen was collected at the left-hand electrode. How would you expect the volume of oxygen collected to compare with the volume of hydrogen collected at the right-hand electrode?

Analyse and interpret data

❻ What ions are present in a solution of sulfuric acid?

❼ Write a half-equation showing the formation of the hydrogen at the right-hand electrode.

SCIENCE IN CONTEXT

ELECTROPLATING

Electroplating involves using electrolysis to coat an object with a thin film of metal. Often this is done for economic reasons, with a fairly cheap metal like steel or nickel being coated with more expensive metals like silver, gold or chromium. Expensive-looking 'silver' knives and forks sometimes have the letters EPNS stamped on them. EPNS stands for **E**lectro-**P**lated **N**ickel **S**ilver. The item is made from nickel with a thin coating of silver added by electrolysis.

Electroplating can also be used to modify the chemical reactivity of the object plated. One example of this is steel cans for food containers plated with a thin layer of tin inside. Tin itself is too soft and expensive to use for the whole can, but it is fairly unreactive and prevents the food from causing the steel to rust.

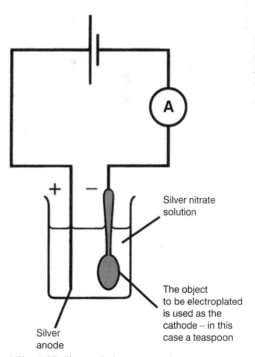

△ Fig. 1.95 Electroplating.

End of topic checklist

An **anode** is the positive electrode.

A **cathode** is the negative electrode.

Electrolysis is the breaking down of a compound by the passage of electricity.

An **electrolyte** is a substance which when molten or dissolved in water allows an electric current to pass through it.

An **anion** is a negative ion that is attracted to the anode.

A **cation** is a positive ion that is attracted to the cathode.

Oxidation is the loss of electrons.

Reduction is the gain of electrons.

The facts and ideas that you should understand by studying this topic:

○ Understand that an electric current is a flow of electrons in a conductor or ions in an electrolyte.

○ Understand that an electrolyte must contain ions and so covalent compounds do not conduct electricity.

○ Understand that for a substance to act as an electrolyte the ions must be free to move.

○ Be able to describe the electrolysis, using inert electrodes, of molten substances, such as lead(II) bromide and predict the products.

○ Be able to describe the electrolysis, using inert electrodes, of the following electrolytes and predict the products formed from aqueous solutions including:

- aqueous sodium chloride solution
- aqueous copper(II) sulfate
- dilute sulfuric acid

○ Be able to write half-equations to represent the reactions that occur at the electrodes.

○ Be able to interpret half-equations as oxidation or reduction.

End of topic questions

1. Explain the following terms:

   a) electrolysis (1 mark)

   b) electrolyte (1 mark)

   c) electrode (1 mark)

   d) anode (1 mark)

   e) cathode. (1 mark)

2. Zinc bromide $ZnBr_2$ is an ionic solid. Why does the solid not conduct electricity? (2 marks)

3. Copy and complete the following table which shows the products formed when a molten electrolyte undergoes electrolysis. (4 marks)

Electrolyte	Product at the anode	Product at the cathode
Silver bromide		
Lead(II) chloride		
Aluminium oxide		
	iodine	magnesium

4. Sodium chloride, NaCl, is ionic. What are the products at the anode and cathode in the electrolysis of:

   a) molten sodium chloride (2 marks)

   b) aqueous sodium chloride? (2 marks)

5. Write half-equations for the following reactions:

   a) The formation of aluminium atoms from aluminium ions (2 marks)

   b) The formation of sodium atoms from sodium ions (2 marks)

   c) The formation of oxygen from oxide ions (2 marks)

   d) The formation of bromine from bromide ions (2 marks)

   e) The formation of oxygen and water from hydroxide ions. (2 marks)

6. An iron fork is to be silver plated using electrolysis. In the electrical circuit:

   a) Which electrode would be the iron fork be? (1 mark)

   b) What would be used as the other electrode? (1 mark)

Exam-style questions
Sample student answer

Question 1

a) The diagram shows the arrangement of particles in the three states of matter.

Each circle represents a particle.

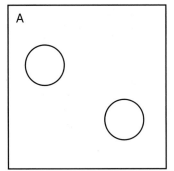

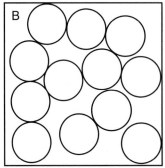

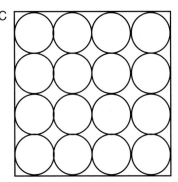

Use the letters A, B and C to give the starting and finishing states of matter for each of the changes in the table.

	Change	Starting state	Finishing state
i)	The formation of water vapour from a puddle of water on a hot day	B	A ✓ ①
ii)	The formation of solid iron from molten iron	B	C ✓ ①
iii)	Melting ice	C	A ✗
iv)	The reaction whose equation is ammonium chloride(s) → ammonia(g) + hydrogen chloride(g)	C	A ✓ ①

(4)

b) Which state of matter is the **least** common for the elements of the Periodic Table at room temperature?

gases ✗

(1)

c) i) Correct – sulfur

ii) Incorrect – this is a 'mixture' of two elements.

iii) Correct – a mixture of an element (O_2) and a compound (H_2O)

iv) Correct – sulfuric acid

The answers rely on using the state symbols for the equation and a thorough knowledge of the terms: element, mixture and compound.

Exam-style questions cont.

c) The manufacture of sulfuric acid can be summarised by the equation

$$2S(s) + 3O_2(g) + 2H_2O(l) \rightarrow 2H_2SO_4(l)$$

Tick one box in each line to show whether the formulae in the table represent a compound, an element or a mixture.

		Compound	Element	Mixture	
i)	$2S(s)$		✓		✓ ①
ii)	$2S(s) + 3O_2(g)$		✓		✗
iii)	$3O_2(g) + 2H_2O(l)$			✓	✓ ①
iv)	$2H_2SO_4(l)$	✓			✓ ①

(4)

(Total 9 marks)

$\frac{6}{9}$

Question 2

This question is about atoms.

a) i) Copy the diagram of an atom. Choose words from the box to label it. (3)

proton	neutron	electron	ion

ii) What is the atomic number of this atom? (1)

iii) What is the mass number of this atom? (1)

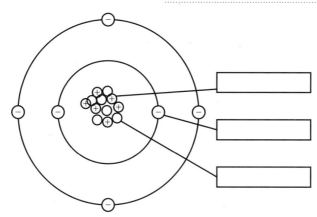

b) Carbon has three isotopes. State one way in which the atoms of the three isotopes are:

  i) the same ....

  ii) different. .... **(2)**

(Total 7 marks)

Question 3

a) Some elements combine together to form ionic compounds. Copy the sentences and use words from the box to complete them.

Each word may be used once, more than once or not at all.

gained	high	lost	low	solid
medium	metals	non-metals	shared	gas

Ionic compounds are formed between .... and ....

Electrons are .... by atoms of one element and .... by atoms of the other element.

The ionic compound formed has a .... melting point and a ....
boiling point. **(6)**

b) Two elements react to form an ionic compound with the formula $MgCl_2$.
(Atomic number of Mg = 12; atomic number of Cl = 17)

  i) Give the electronic configurations of the two elements in this compound **before** the reaction. .... **(2)**

  ii) Give the electronic configurations of the two elements in this compound **after** the reaction. .... **(2)**

(Total 10 marks)

Question 4

9.12 g of iron(II) sulfate was heated. It decomposes to sulfur dioxide ($SO_2(g)$), sulfur trioxide ($SO_3(g)$) and iron(III) oxide. Calculate the mass of iron(III) oxide formed and the volume of sulfur trioxide produced (measured at room temperature and pressure).

(A_r: O = 16, S = 32, Fe = 56; 1 mole of gas at room temperature and pressure occupies 24 000 cm³)

$$2FeSO_4(s) \rightarrow Fe_2O_3(s) + SO_2(g) + SO_3(g)$$

(Total 6 marks)

Exam-style questions cont.

Question 5

The diagram shows the apparatus used to electrolyse lead(II) bromide. (Br = 80)

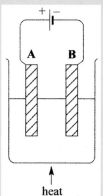

a) The wires connected to the electrodes are made of copper.

Explain why copper conducts electricity. .... (1)

b) Explain why electrolysis does not occur unless the lead(II) bromide is molten. .... (2)

c) The reactions occurring at the electrodes can be represented by the equations shown in the table.

Copy and complete the table to show the electrode (A or B) at which each reaction occurs, and the type of reaction occurring (oxidation or reduction). .... (2)

Electrode reaction	Electrode	Type of reaction
$Pb^{2+} + 2e^- \rightarrow Pb$		
$2Br^- \rightarrow Br_2 + 2e^-$		

(Total 5 marks)

Question 6

On heating calcium carbonate decomposes as shown in the equation:

$$CaCO_3(s) \rightarrow CaO(s) + CO_2(g)$$

5g of calcium carbonate are heated strongly. Calculate:

a) The mass of calcium oxide produced.

b) The volume of carbon dioxide released at room temperature and pressure.

(Relative atomic masses: C =12; O = 16; Ca = 40. The molar volume of a gas = 24 dm³ at rtp.)

(Total 7 marks)

Question 7

Carbon dioxide and graphite have covalently bonded structures. By referring to their structures, explain the following:

a) Carbon dioxide and graphite have very different melting points. .... (8)

b) Graphite conducts electricity but solid carbon dioxide does not. .... (4)

(Total 12 marks)

This section concentrates on a 'branch' of chemistry known as inorganic chemistry. It focuses on the chemical elements, of which there are over 100. This may sound like too much to learn, but the good news is that not all 100 elements are studied! However, because the chemical elements are arranged in a particular pattern, known as the Periodic Table, learning about one element will often provide a very good idea of how other elements will behave.

The previous section included a topic on the Periodic Table. In this section you will study a group of metals and a group of non-metals, followed by oxygen and other gases in the atmosphere. The reactivity series will also be introduced which will provide a good introduction into the extraction of metals. Acids, bases and alkalis will be looked at in some detail as well as the nature and preparation of salts. Finally, this study of the elements and some of their compounds provides an opportunity to look at some of the analytical tests that are used to identify elements and some of their compounds.

STARTING POINTS

1. What is an element – how would you define the term?

2. What does the atomic number of an atom tell you about its structure?

3. In terms of electronic structures, what is the difference between a metal and a non-metal?

4. You will be learning about the Periodic Table of elements. Look at the Periodic Table and make a list of the things that you notice about it.

5. You will be learning about the composition of gases in the air. Do you know, or can you find out, which is the most abundant gas in the air?

6. Make a list of about six to eight metals that you have come across. Which metal in your list do you think is the most reactive? Which metal do you think is the least reactive? Explain your choices.

SECTION CONTENTS

a) Group 1 elements

b) Group 7 elements

c) Gases in the atmosphere

d) Reactivity series

e) Extraction and uses of metals

f) Acids, alkalis and titrations

g) Acids, bases and salt preparations

h) Chemical tests

i) Exam-style questions

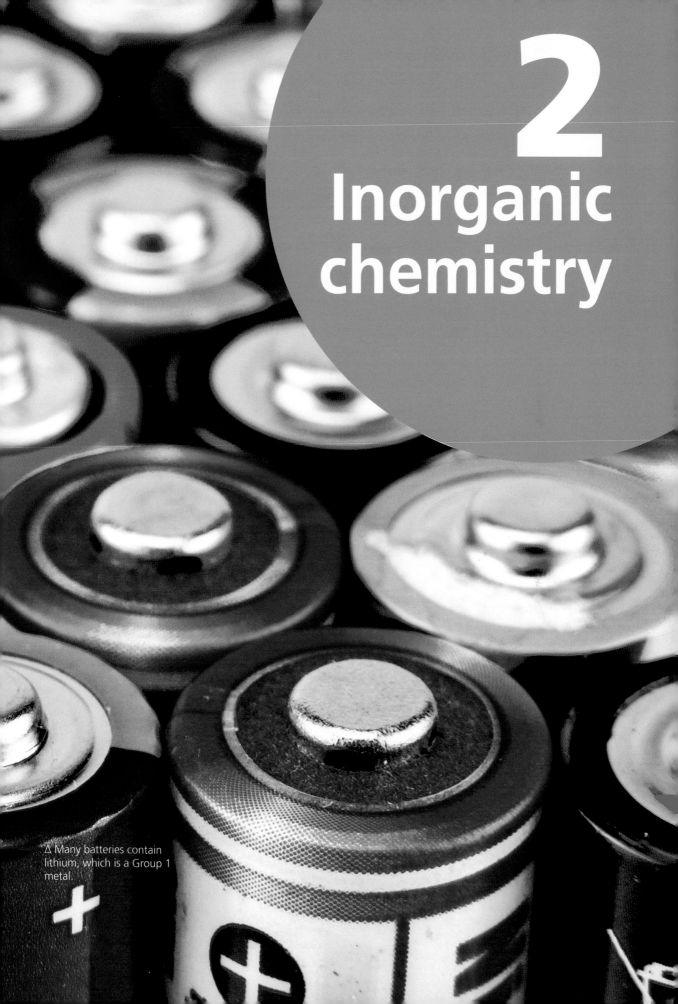

2
Inorganic chemistry

△ Many batteries contain lithium, which is a Group 1 metal.

△ Fig. 2.1 Potassium reacting with water.

Group 1 (alkali metals) – lithium, sodium and potassium

INTRODUCTION

Metals are positioned on the left-hand side and in the middle of the Periodic Table. Therefore the Group 1 elements are metals, but rather different from the metals in everyday use. In fact, when you see how they react with air and water, it is hard to think how they could be used outside the laboratory. This very high reactivity makes them interesting to study. The focus will be on the first three elements in the group: lithium, sodium and potassium. Rubidium, caesium and francium are not available in schools because they are too reactive.

KNOWLEDGE CHECK

✓ Understand that metals are positioned on the left-hand side and middle of the Periodic Table.
✓ Know that elements in a group have similar electron arrangements and have similar chemical properties.
✓ Know that metal oxides are basic and those that dissolve in water form alkalis.

LEARNING OBJECTIVES

✓ Understand how the similarities in the reactions of these elements with water provide evidence for their recognition as a family of elements.
✓ Be able to describe the reactions of the Group 1 elements with air and water and how the differences between these reactions provide evidence for the trend in reactivity in Group 1.
✓ Be able to use knowledge of trends in Group 1 to predict the properties of other alkali metals.
✓ Be able to explain the trend in reactivity in Group 1 in terms of electronic configurations.

REACTIVITY OF THE GROUP 1 ELEMENTS

All the Group 1 elements react with water to produce an alkaline solution. This makes them recognisable as a 'family' of elements, often called the **alkali metals**.

These very reactive metals all have only one electron in their outer electron shell. This electron is easily given away when the metal reacts with non-metals. The more electrons a metal atom has to lose in a reaction, the more energy is needed to start the reaction. This is why the Group 2 elements are less reactive – they have to lose two electrons when they react.

The reactivity of the Group 1 elements depends on the ability of the elements to lose an outermost electron and form a positive ion with a complete outer shell of electrons. The reactivity of the Group 1 elements increases down the group, as the period number increases. This is a result of two factors:

i) As the atom gains more shells of electrons, the outermost electron is further from the attractive force of the nucleus (the protons in the nucleus).

ii) As the atom gains more shells of electrons, the outermost electron is shielded from the attractive force of the nucleus by the inner shells of electrons.

These two factors outweigh the effect of the nucleus having progressively greater positive charge (more protons) as the period number increases.

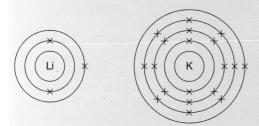

◁ Fig. 2.2 The outer electron in a potassium atom is further away from the attraction of the nucleus than is the case with lithium. Potassium loses its outer electron more easily and so is more reactive.

PROPERTIES OF THE GROUP 1 METALS

The properties of Group 1 metals are as follows:

- Soft to cut.
- Shiny when cut, but quickly tarnishes in air due to the reaction with oxygen and the formation of an oxide.
- Very low melting points compared with most metals.
- Very low densities compared with most metals (lithium, sodium and potassium will float on water).
- React very easily with air, water and elements such as chlorine. The alkali metals are so reactive that they are stored in oil to prevent reaction with air and water.

△ Fig. 2.3 The freshly cut surface of sodium.

Li	lithium	3
Na	sodium	11
K	potassium	19
Rb	rubidium	37
Cs	caesium	55

INCREASING — REACTIVITY

△ Fig. 2.4 Group 1 elements become more reactive as you go further down the group.

QUESTIONS

1. Why are the Group 1 elements known as the alkali metals?

2. How many electrons do the Group 1 elements have in their outer shell?

3. Why is potassium more reactive than lithium?

4. The Group 1 metals are unusual metals. Give one property they have that is different to most other metals.

Reaction	Observations	Equations
Air or oxygen Δ Fig. 2.5	The metals burn easily and their compounds colour flames: • lithium – red • sodium – yellow • potassium – lilac • The tarnishing of the metal to form an oxide and the ease of burning increase from lithium to potassium. A white solid oxide is formed in each case.	lithium + oxygen → lithium oxide $4Li(s) + O_2(g) \rightarrow 2Li_2O(s)$ sodium + oxygen → sodium oxide $4Na(s) + O_2(g) \rightarrow 2Na_2O(s)$ potassium + oxygen → potassium oxide $4K(s) + O_2(g) \rightarrow 2K_2O(s)$
Water Δ Fig. 2.6	The metals react vigorously. They float on the surface, moving around rapidly. With both sodium and potassium the heat of the reaction melts the metal so it forms a sphere. Bubbles of gas are given off, and the metal 'disappears'. With the more reactive metals (such as potassium) the hydrogen gas produced burns. The resulting solution is alkaline. The rate of bubbling or effervescence increases dramatically from lithium to potassium.	lithium + water → lithium hydroxide + hydrogen $2Li(s) + 2H_2O(l) \rightarrow 2LiOH(aq) + H_2(g)$ sodium + water → sodium hydroxide + hydrogen $2Na(s) + 2H_2O(l) \rightarrow 2NaOH(aq) + H_2(g)$ potassium + water → potassium hydroxide + hydrogen $2K(s) + 2H_2O(l) \rightarrow 2KOH(aq) + H_2(g)$

Δ Table 2.1 Reactions of Group 1 metals.

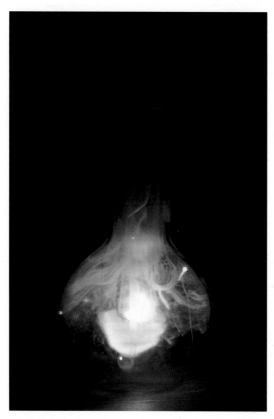

△ Fig. 2.7 Sodium burning in chlorine.

QUESTIONS

1. What colour is sodium oxide, formed when sodium is burned in oxygen?

2. What gas is produced when potassium reacts with water? What is the name of the solution formed in this reaction?

3. A student has written: 'Sodium is a more reactive element than lithium.' Can you justify this statement?

FACTS ABOUT THE GROUP 1 METALS

1. Lithium is found in large quantities (estimated at 230 billion tonnes) in compounds in seawater.

2. Sodium is found in many minerals and is the sixth most abundant element overall in the Earth's crust (amounting to 2.6% by weight).

3. Potassium is also found in many minerals and is the seventh most abundant element in the Earth's crust (amounting to 1.5% by weight).

4. Rubidium was discovered by Bunsen (of Bunsen burner fame) in 1861. It is more abundant than copper, about the same as zinc, and is found in very small quantities in a large number of minerals. Because of this low concentration in mineral deposits, only 2 to 4 tonnes of rubidium are produced each year worldwide.

5. Caesium is more abundant than tin, mercury and silver. However, its very high reactivity makes it very difficult to extract from mineral deposits.

6. Francium was discovered as recently as 1939 as a product of the radioactive decay of an isotope of actinium.

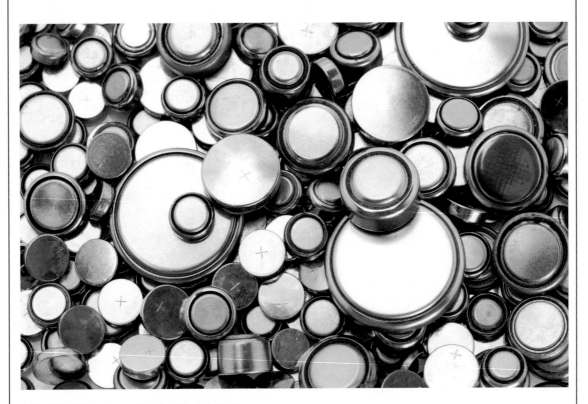

△ Fig. 2.8 Lithium is used in all of these batteries.

End of topic checklist

An **alkali metal** is a metal in Group 1 of the Periodic Table.

The facts and ideas that you should know and understand by studying this topic:

○ Know that lithium, sodium and potassium have similar reactions with water and this is why they grouped together in Group 1 of the Periodic Table.

○ Be able to describe the reactions of lithium, sodium and potassium with air and water.

○ Know that their reactions with water produce an alkali and hydrogen.

○ Know that the elements become increasingly reactive as you go down the group, from lithium to potassium, and that this trend allows the properties of other alkali metals such as francium to be predicted.

○ Understand that the reactivity of the elements depends on the distance between the outer electron and the nucleus: the further the distance of the outer electron from the nucleus, the more reactive the metal is.

End of topic questions

1. This question is about the Group 1 elements: lithium, sodium and potassium.

   a) Which is the most reactive of the elements? **(1 mark)**

   b) Why are the elements stored in oil? **(1 mark)**

   c) Why do the elements tarnish quickly when they are cut? **(1 mark)**

   d) Why does sodium float when added to water? **(1 mark)**

2. Why are the Group 1 elements known as the alkali metals? **(2 marks)**

3. Write word equations and balanced symbol equations for the following reactions:

   a) lithium and oxygen **(3 marks)**

   b) potassium and water **(3 marks)**

4. How do the reactions of the alkali metals with water provide evidence for the trend of reactivity in Group 1? **(2 marks)**

5. Explain why potassium is more reactive than sodium. **(3 marks)**

6. This question is about rubidium (symbol Rb), which is a less common Group 1 element.

   When rubidium is added to water:

   a) Which gas would be formed? **(1 mark)**

   b) What chemical compound would be formed in solution? What result would you predict if universal indicator solution was added to the solution? **(2 marks)**

   c) Would you expect rubidium to be more or less reactive than potassium? Explain your answer. **(2 marks)**

7. Samples of lithium, sodium and potassium are removed from the oil they are stored in, dried and cut to expose the inner metal surface.

   Which metal surface would you expect to tarnish most rapidly? Explain your answer. **(2 marks)**

Group 7 (halogens) – chlorine, bromine and iodine

INTRODUCTION

Group 7 elements are located on the right-hand side of the Periodic Table with the other non-metals. They look very different from each other, so it may seem strange that they are in the same group. However, their chemical properties are very similar, and all of them are highly reactive. The focus in this topic will be on chlorine, bromine and iodine. Fluorine is a highly reactive gas and astatine is a radioactive black solid with a very short half-life (so will only exist in very small quantities).

△ Fig. 2.9 At room temperature and atmospheric pressure, chlorine is a pale green gas, bromine a red-brown liquid and iodine is a black solid.

KNOWLEDGE CHECK

✓ Understand that non-metals are positioned on the right-hand side of the Periodic Table.
✓ Know that elements in a group have similar electron arrangements and have similar chemical properties.
✓ Be familiar with the terms oxidation and reduction.

LEARNING OBJECTIVES

✓ Be able to describe the colours, physical states (at room temperature) and trends in physical properties of the Group 7 elements.
✓ Know how the reactivity of chlorine, bromine and iodine changes down the group.
✓ Explain the trend in reactivity in Group 7 in terms of electronic configurations.
✓ Know how the trends in physical properties and reactivity of chlorine, bromine and iodine can be used to predict the properties of other halogens.
✓ Understand how displacement reactions involving halogens and halides provide evidence for the trend in reactivity in Group 7.

REACTIVITY OF THE GROUP 7 ELEMENTS

The Group 7 elements are sometimes referred to as the **halogen** elements or halogens.

'Halogen' means 'salt-maker', and halogens react with most metals to make salts.

Halogens have seven electrons in their outermost electron shell, so they only need to gain one electron to obtain a full outer electron shell. This is what makes them very reactive. They react with metals, gaining an electron and forming a singly charged negative ion (see ionic bonding on page 72).

The reactivity of the Group VII elements depends on the ability of the element to gain an electron and form a negative ion with a complete outer electron shell. The reactivity of the Group VII elements decreases down the group, as the period number increases. This is a result of two factors:

i) As the atom gains more shells of electrons, an incoming electron is further from the attractive force of the nucleus (the protons in the nucleus).

ii) As the atom gains more shells of electrons, the incoming electron is shielded from the attractive force of the nucleus by the inner shells of electrons.

These two factors outweigh the effect of the nucleus having progressively greater positive charge (more protons) as the period number increases.

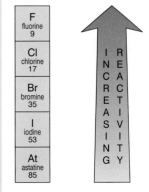

△ Fig. 2.10 Increasing reactivity goes up Group 7.

PROPERTIES OF THE GROUP 7 ELEMENTS

The properties of the Group 7 elements are as follows.

- Appearance at room temperature: fluorine is a pale yellow gas; chlorine is a pale green gas; bromine is a red-brown liquid; iodine is a black shiny solid.
- The melting and boiling points of the elements increase down the group as the relative atomic masses increase.
- All have *seven electrons* in their outermost electron shell.
- All exist as diatomic molecules, that is each molecule contains two atoms. For example, F_2, Cl_2, Br_2, I_2.
- Halogens react with water and react with metals to form salts. Iodine has very low solubility and little reaction with water.
- They undergo **displacement reactions**.

Halogen	Melting point (°C)	Boiling point (°C)
fluorine	−220	−188
chlorine	−101	−35
bromine	−7.2	58.8
iodine	114	184
astatine	302	337

△ Table 2.2 Melting and boiling points for the halogens.

Reaction	Observations	Equations
Water chlorine gas water △ Fig. 2.11	The halogens dissolve in water and also react with it, forming solutions that behave as bleaches. Chlorine solution is pale yellow. Bromine solution is orange. Iodine solution is brown.	chlorine + water → hydrochloric acid + chloric(I) acid $Cl_2(g) + H_2O(l) \rightarrow HCl(aq) + HClO(aq)$
Metals chlorine iron wool △ Fig. 2.12	The halogens will form salts with all metals. For example, gold leaf will catch fire in chlorine without heating. With a metal such as iron, brown fumes of iron(III) chloride form.	iron + chlorine → iron(III) chloride $2Fe(s) + 3Cl_2(g) \rightarrow 2FeCl_3(s)$ Fluor*ine* forms salts called fluor*ides*. Chlor*ine* forms salts called chlor*ides*. Brom*ine* forms salts called brom*ides*. Iod*ine* forms salts called iod*ides*.
Displacement chlorine gas potassium iodide solution iodine being formed △ Fig. 2.13	A more reactive halogen will displace a less reactive halogen from a solution of a salt. Chlorine displaces bromine from sodium bromide solution. The colourless solution (sodium bromide) will turn orange as the chlorine is added due to the formation of bromine. Chlorine displaces iodine from sodium iodide solution. The colourless solution (sodium iodide) will turn brown as the chlorine is added due to the formation of iodine.	chlorine + sodium bromide → sodium chloride + bromine $Cl_2(g) + 2NaBr(aq) \rightarrow 2NaCl(aq) + Br_2(aq)$ chlorine + sodium iodide → sodium chloride + iodine $Cl_2(g) + 2NaI(aq) \rightarrow 2NaCl(aq) + I_2(aq)$

△ Table 2.3 Reactions of Group 7 elements.

The displacement reactions between halogens and solutions of halide ions (shown above) are examples of **redox reactions** – that is, oxidation and reduction reactions. The Na^+ ions are 'spectator' ions and play no part in the reaction, so if the reaction between chlorine and sodium bromide solution is written with them removed, the equation becomes:

$$Cl_2(g) + 2Br^-(aq) \rightarrow 2Cl^-(aq) + Br_2(aq)$$

If this reaction is then written as two half-equations showing how electrons are involved:

$Cl_2(g) + 2e^- \rightarrow 2Cl^-(aq)$ The chlorine gas has gained electrons – it has been reduced.

$2Br^-(aq) \rightarrow Br_2(aq) + 2e^-$ The bromide ions have lost electrons – they have been oxidised.

In this reaction the chlorine is an oxidising agent – it oxidises the bromide ions and is itself reduced. The bromide ions are reducing agents – they reduce the chlorine and are themselves oxidised.

More generally, the more reactive halogen will displace the less reactive halogen from one of its compounds.

QUESTIONS

1. What is the trend in melting point from chlorine to iodine?

2. Would you expect astatine to be a solid, liquid or gas at room temperature? Explain your answer.

3. How many electrons do the Group 7 elements have in their outer shell?

4. Why are the Group 7 elements particularly reactive when compared to other non-metals?

5. Chlorine exists as a diatomic molecule. Explain what this means.

6. In the context of the Group 7 elements explain what you understand by the phrase *displacement reaction*.

7. Why are displacement reactions also redox reactions?

Developing investigative skills

A student was provided with three aqueous solutions containing chlorine, bromine and iodine. The student added a few drops of cyclohexane to each solution in separate test tubes and then stirred each with a clean glass rod. The cyclohexane is used as chlorine, bromine and iodine dissolve easily in the cyclohexane and produce distinctive colours. When the cyclohexane layer had separated from the solution, the student recorded the colours of the cyclohexane layers in the table below:

Solution in water	Colour of the cyclohexane layer
chlorine	colourless
bromine	orange
iodine	violet

Note: chlorine water is not currently classified as hazardous, but the chlorine gas that escapes from it is toxic.

The student cleaned out the tubes and then performed a series of test tube reactions as indicated in the table. In each case the student mixed small quantities of solution A with twice the volume of solution B, added 10 drops of cyclohexane and then stirred the mixture with a clean glass rod. Once the cyclohexane layer had separated, the student recorded the results.

Solution A	Solution B	Colour of cyclohexane layer
1. aqueous chlorine	sodium bromide	orange
2. aqueous chlorine	sodium iodide	violet
3. aqueous bromine	sodium chloride	orange
4. aqueous bromine	sodium iodide	violet
5. aqueous iodine	sodium chloride	violet
6. aqueous iodine	sodium bromide	orange

Demonstrate and describe techniques

❶ Cyclohexane is highly flammable and a skin irritant and highly toxic to aquatic organisms and the chlorine and bromine solutions are both irritants. What precautions should the student have taken when doing this experiment?

Analyse and interpret data

❷ What can you deduce about the relative reactivity of chlorine, bromine and iodine from the *first two* results?

❸ Write an equation for the reaction indicated by result 4.

Evaluate data and methods

❹ The student has made a mistake in recording one of the results. Which one? Explain how you know.

SCIENCE IN CONTEXT   **FLUORINE**

Fluorine is the most reactive non-metal in the Periodic Table. It reacts with most other elements except helium, neon and argon. These reactions are often sudden or explosive. Even radon, the very unreactive noble gas, burns with a bright flame in a jet of fluorine gas. All metals react with fluorine to form fluorides. The reactions of fluorine with Group 1 metals are explosive.

Early scientists tried to make fluorine from hydrofluoric acid (HF(aq)) but this proved to be highly dangerous, killing or blinding several scientists who attempted it. They became known as the 'fluorine martyrs'. Today fluorine is manufactured by the electrolysis of the mineral fluorite, which is calcium fluoride.

Fluorine is not an element to meddle with. You will certainly not see it in your laboratory!

EXTENSION

Although the halogens are potentially harmful their properties make them very useful in our lives. Use your knowledge of atomic structure, bonding and reaction types to answer the questions.

1. Chlorine is a pale green gas which can be obtained by the electrolysis of an aqueous solution of sodium chloride. Chlorine can be used to kill bacteria and is used in the manufacture of bleach.

   a) The electronic structure of a chlorine atom is 2,8,7. Draw simple diagrams to show the arrangement of the outer electrons in a diatomic molecule of Cl_2 and a chloride ion.

   b) In the electrolysis of an aqueous solution of sodium chloride the positive anode attracts the OH^- ions and Cl^- ions to form chlorine molecules. Copy and complete the equation below and explain why this is oxidation.

   _____$Cl^-(aq) \rightarrow$ _____$(g) + 2e^-$

   c) Sodium hydroxide solution and chlorine will react at room temperature to form bleach, sodium chlorate (NaClO) which is used in domestic cleaning agents. Copy and complete and balance the equation shown below.

   sodium hydroxide + chlorine $\rightarrow$ sodium chloride + sodium chlorate(I) + water

   _____ $+ Cl_2(aq) \rightarrow$ _____ $+ NaClO(aq)$ $+ H_2O(l)$

d) Chlorine will displace bromine from a solution of potassium bromide to form bromine and potassium chloride. Explain why this reaction takes place and describe what you would observe if chlorine water was added to a solution of potassium bromide in a test tube.

2. Fluorine is a pale yellow gas and is the most reactive of the chemical elements. It is so reactive that glass, metals and even water burn with a bright flame in a jet of fluorine gas. Fluorides however are often added to toothpaste and, controversially, to some water supplies to prevent dental cavities.

a) Give a reason why fluorine is so reactive.

b) Write an ionic half-equation which shows the conversion of fluorine to fluoride ions.

c) Potassium fluoride is a compound which may be found in toothpaste. Explain why fluorine cannot be displaced from this compound using either chlorine or iodine.

End of topic checklist

A **halogen** is the name given to an element in Group 7 of the Periodic Table.

A **displacement reaction** is one in which one element takes the place of another element in a compound and removes (displaces) it from the compound.

The facts and ideas that you should know and understand by studying this topic:

○ Know the colours and physical states of chlorine, bromine and iodine at room temperature.

○ Be able to describe the trend in physical properties of the halogens.

○ Be able to describe the relative reactivity of the Group 7 elements.

○ Be able to predict the likely properties of fluorine and astatine.

○ Be able to describe and explain displacement reactions involving the halogens and solutions of their compounds.

○ Understand that these displacement reactions are redox reactions and involve oxidation and reduction.

○ Be able to explain the trend in reactivity in Group 7 in terms of electronic configurations.

End of topic questions

1. This question is about the Group 7 elements: chlorine, bromine and iodine.

   a) Which is the most reactive of the elements? (1 mark)

   b) Which of the elements exists as a liquid at room temperature and pressure? (1 mark)

   c) Which of the elements exists as a solid at room temperature and pressure? (1 mark)

   d) What is the appearance of bromine? (1 mark)

2. Explain the following:

   a) The Group 7 elements are the most reactive non-metals. (2 marks)

   b) The most reactive element is at the top of the group. (2 marks)

3. Write word and balanced symbol equations for the following reactions:

   a) sodium and chlorine (3 marks)

   b) magnesium and bromine (3 marks)

   c) hydrogen and fluorine. (3 marks)

4. Aqueous bromine reacts with sodium iodide solution.

   a) What type of chemical reaction is this? (1 mark)

   b) Write a balanced symbol equation for this reaction. (2 marks)

   c) This reaction involves oxidation and reduction:

      i) What is oxidation? What has been oxidised in this reaction? (2 marks)

      ii) What is reduction? What has been reduced in this reaction? (2 marks)

Gases in the atmosphere

△ Fig. 2.14 Magnesium burning in oxygen.

INTRODUCTION

Oxygen is arguably the most important of all gases. Without it, animals would not survive and many familiar everyday processes would not take place. However, oxygen supports combustion – if the air around us was made up of just oxygen, fuels would burn very quickly! Some of the non-metallic oxides it forms are pollutants and can cause serious damage to the environment. Oxygen also reacts with many of the elements in the Periodic Table to form **oxides**.

KNOWLEDGE CHECK

✓ Know that oxygen and carbon dioxide are important gases present in the air.
✓ Know that metals form basic oxides and non-metals usually form acidic oxides.
✓ Know that oxygen gas exists as a diatomic molecule, O_2.

LEARNING OBJECTIVES

✓ Know the approximate proportions by volume of the four main gases in dry air.
✓ Be able to describe experiments to determine the percentage by volume of oxygen in air, involving the reaction of a metal (e.g. iron) or non-metal (e.g. phosphorus) with air.
✓ Describe the combustion of elements in oxygen, including magnesium, hydrogen and sulfur.
✓ Be able to describe the formation of carbon dioxide from the thermal decomposition of metal carbonates, including copper(II) carbonate.
✓ Understand that carbon dioxide is a greenhouse gas and may contribute to climate change.

THE GASES IN THE AIR

These are the main gases found in normal, dry air and their approximate percentage by volume. Air also contains small quantities of the other noble gases – neon, helium, krypton and xenon – and it may contain water vapour.

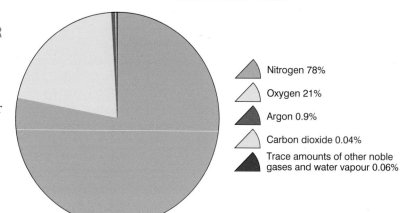

- Nitrogen 78%
- Oxygen 21%
- Argon 0.9%
- Carbon dioxide 0.04%
- Trace amounts of other noble gases and water vapour 0.06%

△ Fig. 2.15 Gases in the air by volume

TO DETERMINE THE PERCENTAGE BY VOLUME OF OXYGEN IN THE AIR

Here is a simple experiment to show the percentage of oxygen in the air.

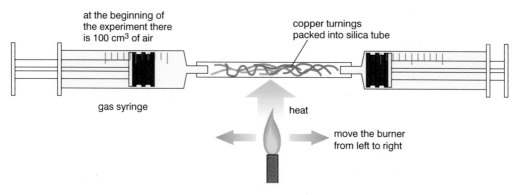

△ Fig. 2.16 Experiment to show the percentage of oxygen in air.

At the beginning the syringes contain 100 cm³ of air. The gas is continually passing from one syringe to the other over the copper in the silica tube. When the copper is heated, the amount of gas gradually decreases as the oxygen in the air is used up. The copper reacts with the oxygen to form copper(II) oxide.

copper	+	oxygen	→	copper(II) oxide
$2Cu(s)$	+	$O_2(g)$	→	$2\ CuO(s)$

Eventually the gas volume stops contracting because all the oxygen has been used up. The volume of gas in the syringes goes down to about 79 cm³, showing that about 21 cm³ of the air was oxygen: that is, 21%.

Early chemists used phosphorus to show that oxygen formed about one-fifth of the air. They used phosphorus in a bell jar of air as shown in the diagram. The phosphorus reacted with the oxygen in the air and formed phosphorus(V) oxide, which dissolved in the water. The water rose steadily up the bell jar to take the place of the oxygen.

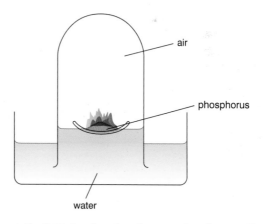

△ Fig. 2.17 Burning phosphorus to show how much oxygen is in air.

phosphorus	+	oxygen	→	phosphorus(V) oxide
$4P(s)$	+	$5O_2(g)$	→	$2P_2O_5(s)$

1. Which noble gas is found in the air in the greatest quantity?
2. What is the percentage of carbon dioxide in normal air?
3. Which is the most abundant gas in the air?
4. What chemical is produced when air is passed over hot copper?

THE REACTIONS OF OXYGEN IN THE AIR

The oxides of elements can often be made by heating the element in air or oxygen. This is a **combustion reaction**. For example, the metal magnesium burns in oxygen to form magnesium oxide:

magnesium	+	oxygen	→	magnesium oxide
$2Mg(s)$	+	$O_2(g)$	→	$2MgO(s)$

Magnesium oxide forms as a white powder. When distilled water is added to the powder and the mixture is tested with universal indicator, the pH is greater than 7, showing that the oxide has formed an alkaline solution.

Hydrogen burns in air to form water.

hydrogen	+	oxygen	→	water
$2H_2(g)$	+	$O_2(g)$	→	$2H_2O(l)$

When sulfur is burned in oxygen, sulfur dioxide gas is formed:

sulfur	+	oxygen	→	sulfur dioxide
$S(s)$	+	$O_2(g)$	→	$SO_2(g)$

When this is dissolved in water and then tested with universal indicator solution, the pH is less than 7, showing that the oxide has formed an acidic solution.

If it is in powder or wire form, iron will burn in air and react with oxygen. An orange oxide forms.

iron	+	oxygen	→	iron(III) oxide
$4Fe(s)$	+	$3O_2(g)$	→	$2Fe_2O_3(s)$

△ Fig. 2.18 Sulfur burns in oxygen with a blue flame to give a colourless gas.

Copper reacts slowly with oxygen. It turns gradually darker on heating and also makes a black oxide.

copper	+	oxygen	→	copper(II) oxide
$2Cu(s)$	+	$O_2(g)$	→	$2CuO(s)$

All the reactions of elements with oxygen are oxidation reactions. The metal oxides shown in the reactions above are *basic* oxides (they react with acids). The non-metal oxides shown are *acidic* oxides (they react with bases or alkalis). Some non-metal oxides are *neutral* oxides (they have no basic or acidic properties).

THE HINDENBURG DISASTER

In the 1930s, rigid airships called zeppelins regularly crossed the Atlantic Ocean between Germany and the USA. The Hindenburg, a zeppelin filled with hydrogen, burst into flames in New Jersey in the USA on 6 May 1937, killing 36 people and ending this form of travel.

Knowing how flammable hydrogen is, you might be surprised that such a form of transport was ever considered safe, but German commercial zeppelins like the Hindenburg had been in use for more than

Δ Fig. 2.19 The airship, the Hindenburg, was full of hydrogen. It caught fire while trying to dock in the USA.

30 years. During this time tens of thousands of passengers flew over a million miles on more than 2000 flights without a single injury. Some of the zeppelins were filled with helium rather than hydrogen – a much safer choice. The cause of the Hindenburg disaster has never been confirmed, but it is fairly clear that something, possibly a spark, set the hydrogen alight.

For a number of years hydrogen has been talked about as the ideal fuel for motor cars. Surely a fuel that burns to form only water has to be a winner? The problem is that making hydrogen produces a large quantity of carbon dioxide and therefore, overall, it is still a polluting fuel. Then there is the question of safety – what if there was a problem when filling up the car and hydrogen leaked out into the atmosphere…?

QUESTIONS

1. What is a basic oxide?

2. Calcium is a Group 2 metal that reacts with oxygen. Write a fully balanced equation, including state symbols, for the reaction of calcium with oxygen.

3. Would you expect phosphorus(V) oxide to be a basic or acidic oxide? Explain your answer.

Developing investigative skills

A student set up the apparatus as shown in the diagram with the long tube turned upside-down in a trough of water. Previously some iron filings had been sprinkled into the tube, and many of these had stuck to the side of the tube. With the same levels of water in the tube and the trough, the student recorded the volume of air in the tube (100 cm³).

After about three hours the student returned to the apparatus, equalised the water levels as before and took a second reading of the volume of air in the tube (85 cm³). The student then worked out how much of the air had been replaced by water. (Note: iron filings can cause severe irritation in eyes.)

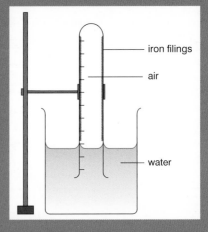

△ Fig. 2.20 Apparatus for experiment.

Make observations and measurements

❶ Draw a table and record the student's results.

❷ Describe how you would expect the iron filings to change during the experiment.

Analyse and interpret data

❸ What chemical reaction was taking place in the tube? Write a balanced equation for the reaction.

❹ What do the results indicate about the composition of the air in the tube?

Evaluate data and methods

❺ The results in this experiment are different to those obtained in the reaction between copper and air on page 127. Suggest some possible reasons for this.

❻ Why did the student equalise the water levels in the trough and the tube before taking the reading of the volume of gas in the tube?

CARBON DIOXIDE

Carbon dioxide is an important gas. It is formed in large quantities when fossil fuels are burnt.

▷ Fig. 2.21 Charcoal is mainly carbon. It burns to give carbon dioxide.

Carbon dioxide can be prepared by the **thermal decomposition** of certain metal carbonates. Copper(II) carbonate and zinc carbonate are examples:

copper(II) carbonate	$\rightarrow$	copper(II) oxide	+	carbon dioxide
$CuCO_3(s)$	$\rightarrow$	$CuO(s)$	+	$CO_2(g)$
green		black		

zinc carbonate	$\rightarrow$	zinc oxide	+	carbon dioxide
$ZnCO_3(s)$	$\rightarrow$	$ZnO(s)$	+	$CO_2(g)$
white		white (yellow when hot)		

Carbon dioxide is often referred to as a **greenhouse gas**. Some scientists think that the increase in the amount of carbon dioxide in the atmosphere, as a result of burning fossil fuels and fuels made from fossil fuels such as petrol, is contributing to climate change or global warming. Other scientists believe that recent changes in climatic conditions are just part of natural changes that have occurred since the Earth was formed.

THE MANUFACTURE OF OXYGEN AND NITROGEN

Oxygen, nitrogen and the noble gases can all be manufactured from the air. The industrial process has several stages:

1. The air is cooled to about −80 °C so that carbon dioxide and water vapour solidify and can be removed.

2. The air is cooled further, compressed and then allowed to expand quickly. This causes further cooling, and at about −200 °C the air becomes a liquid.

3. The liquid air is fractionally distilled. This involves using a large fractionating column which separates liquids with different boiling points. Oxygen's boiling point is −183 °C and nitrogen's is −196 °C: they can be separated.

Oxygen and nitrogen are stored in large metal cylinders under high pressure and so remain as liquids until the pressure is released. Oxygen is used in medicine to support patients with breathing difficulties. It is also mixed with hydrocarbons such as acetylene (ethyne) for use in cutting tools. Nitrogen's uses depend on its very low reactivity. It is used in food packaging as a way of excluding air and stopping oxidation and decay. It is also often used to reduce fire hazards, particularly in military aircraft fuel systems.

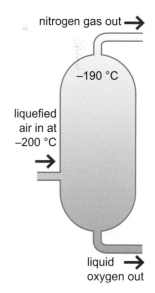

△ Fig. 2.22 Fractional distillation of liquid air.

End of topic checklist

Combustion is a process involving the burning of a substance in air or oxygen.

Thermal decomposition is the breaking down of a compound by heat.

The facts and ideas that you should know and understand by studying this topic:

○ Know that the air contains approximately 78% nitrogen, 21% oxygen, 0.9% argon and 0.04% carbon dioxide.

○ Be able to explain laboratory experiments that can be used to determine the percentage of oxygen by volume in the air (copper, iron and phosphorus reacting with oxygen).

○ Be able to describe the combustion reactions of oxygen with magnesium, hydrogen and sulfur, and to write equations for these reactions.

○ Be able to describe the formation of carbon dioxide from the thermal decomposition of copper(II) carbonate.

○ Understand that carbon dioxide is a greenhouse gas and may contribute to climate change.

End of topic questions

1. This question is about the composition of a sample of normal air.

   a) What is the proportion of oxygen? **(1 mark)**

   b) What is the proportion of carbon dioxide? **(1 mark)**

2. Explain the following:

   a) When air is passed over heated copper in a long glass tube the volume of the air decreases. **(2 marks)**

   b) When iron wool is put at the bottom of a test tube which is then inverted in a trough of water, the water rises up the tube. **(2 marks)**

3. **a)** Write fully balanced symbol equations for the reactions of oxygen with:

   i) sodium **(2 marks)**

   ii) aluminium **(2 marks)**

   iii) magnesium. **(2 marks)**

   b) Are these oxides acidic or basic? Explain your answer. **(1 mark)**

4. Write a balanced symbol equation for the combustion of hydrogen in air. **(2 marks)**

5. Carbon dioxide can also be made by the decomposition of copper(II) carbonate.

   a) What does the term decomposition mean? **(1 mark)**

   b) Write a balanced symbol equation for the reaction. **(2 marks)**

6. Carbon dioxide is said to be a 'greenhouse gas'. What does this mean? **(2 marks)**

7. Not all scientists are convinced that there is such a thing as the greenhouse effect. What is the greenhouse effect and what is another possible explanation for climate change? **(4 marks)**

Reactivity series

INTRODUCTION

The Periodic Table is a way of ordering the chemical elements that highlights their similar and different properties. The **reactivity series** is another way of classifying elements, this time in order of their reactivity to help explain or predict their reactions. This has many practical applications, such as being able to predict how metals can be extracted from their ores and how the negative effects of the chemical process of rusting can be reduced.

Δ Fig. 2.23 This ship is made of steel, which rusts.

KNOWLEDGE CHECK

✓ Know that metals have different reactivity.
✓ Understand what a displacement reaction is.
✓ Understand what oxidation and reduction are.

LEARNING OBJECTIVES

✓ Understand how metals can be arranged in a reactivity series based on their reactions with water, dilute hydrochloric or sulfuric acid.
✓ Understand how metals can be arranged in a reactivity series based on their displacement reactions between metals and metal oxides, or metals and aqueous solutions of metal salts.
✓ Know the order of reactivity of these metals: potassium, sodium, lithium, calcium, magnesium, aluminium, zinc, iron, copper, silver and gold.
✓ Know the conditions under which iron rusts.
✓ Understand how the rusting of iron may be prevented by barrier methods, galvanising and sacrificial protection.
✓ Understand the terms oxidation, reduction, redox, oxidising agent and reducing agent in terms of gain or loss of oxygen and loss or gain of electrons.
✓ Be able to investigate reactions between dilute hydrochloric and sulfuric acids and metals (for example, magnesium, zinc and iron).

THE ORDER OF CHEMICAL REACTIVITY

Elements can be arranged in order of their reactivity. The more reactive a metal is, the easier it is to form compounds and the harder it is to break the compounds down. We can predict how metals might react by looking at the reactivity series.

CHEMICAL REACTIONS AND A REACTIVITY SERIES

The pattern in the reactions of a metal can be related to the reactivity series. The most reactive metals react with water at room temperature. For example, potassium, sodium and lithium in Group 1 and calcium in Group 2 of the Periodic Table react rapidly with water.

sodium	+	water	→	sodium hydroxide	+	hydrogen
$2Na(s)$	+	$2H_2O(l)$	→	$2NaOH(aq)$	+	$H_2(g)$

The less reactive metals such as magnesium and iron react with steam.

magnesium	+	steam	→	magnesium oxide	+	hydrogen
$Mg(s)$	+	$H_2O(g)$	→	$MgO(s)$	+	$H_2(g)$

Some of the mid-reactivity metals produce hydrogen when they react with dilute acids. So, for example, magnesium, aluminium, zinc and iron all form hydrogen when they react with dilute hydrochloric acid.

zinc	+	hydrochloric acid	→	zinc chloride	+	hydrogen
$Zn(s)$	+	$2HCl(aq)$	→	$ZnCl_2(aq)$	+	$H_2(g)$

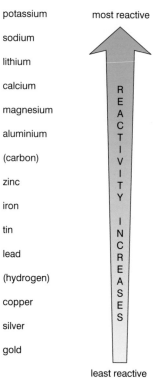

potassium

sodium

lithium

calcium

magnesium

aluminium

(carbon)

zinc

iron

tin

lead

(hydrogen)

copper

silver

gold

most reactive

REACTIVITY INCREASES

least reactive

Δ Fig. 2.24 The reactivity series shows elements, mainly metals, in order of decreasing reactivity.

The metals below hydrogen in the reactivity series do not react to form hydrogen with water or dilute acids.

Another use of the reactivity series is to predict how metals can be extracted from their ores (page 143). The metals that are below carbon in the reactivity series can be obtained by heating their oxides with carbon.

This type of reaction is called a **displacement reaction**. A more reactive element, such as carbon, 'pushes' (or displaces) a less reactive metal, such as iron, out of its compound.

In fact, any element higher up the reactivity series can displace an element lower down the series.

For example, magnesium is higher up the reactivity series than copper. So if magnesium powder is heated with copper(II) oxide, then copper and magnesium oxide are produced.

magnesium	+	copper(II) oxide	→	magnesium oxide	+	copper
$Mg(s)$	+	$CuO(s)$	→	$MgO(s)$	+	$Cu(s)$

This reaction is an example of a **redox reaction**. The magnesium has been oxidised to magnesium oxide and the copper(II) oxide has been reduced to copper. As the magnesium is responsible for the reduction of the copper(II) oxide, it is acting as a **reducing agent**. Similarly, as the copper(II) oxide is responsible for the oxidation of the magnesium, it is acting as an **oxidising agent**. In a redox reaction, the reducing agent is always oxidised and the oxidising agent is always reduced.

What will happen if copper is heated with magnesium oxide? Nothing happens, because copper is lower in the reactivity series than magnesium.

USING DISPLACEMENT REACTIONS TO ESTABLISH A REACTIVITY SERIES

Displacement reactions of metals and their compounds in aqueous solution can be used to work out the reactivity series.

In the same way that a more reactive element can push a less reactive element out of a compound, a more reactive metal ion in aqueous solution can displace a less reactive one.

For example, if you add zinc to copper(II) sulfate solution, the zinc displaces the copper because zinc is more reactive than copper. When the experiment is carried out, the blue colour of the copper ion will fade as copper is produced and zinc ions are made.

zinc	+	copper(II) sulfate solution	$\rightarrow$	zinc sulfate solution	+	copper
$Zn(s)$	+	$Cu^{2+}(aq) + SO_4^{2-}(aq)$	$\rightarrow$	$Zn^{2+}(aq) + SO_4^{2-}(aq)$	+	$Cu(s)$

The zinc atoms have lost electrons to form zinc ions (Zn^{2+}) and have been oxidised. The copper ions (Cu^{2+}) have gained electrons and have been reduced.

To build up a whole reactivity series, a set of reactions can be tried to see whether metals can displace metal ions. They follow the general rule that a more reactive metal can displace a less reactive metal.

For example, you may have seen the reaction of copper wire with silver nitrate solution. As the reaction proceeds, a shiny grey precipitate appears (this is silver) and the solution begins to turn blue as Cu^{2+} ions are produced from the copper.

copper	+	silver nitrate	$\rightarrow$	copper(II) nitrate	+	silver
$Cu(s)$	+	$2Ag^+(aq) + 2NO_3^-(aq)$	$\rightarrow$	$Cu^{2+}(aq) + 2NO_3^-(aq)$	+	$2Ag(s)$

In this reaction the copper atoms lose electrons and the silver ions gain electrons. This shows that silver can be displaced by copper, and so silver is below copper in the reactivity series.

QUESTIONS

1. Will copper react with dilute hydrochloric acid to produce hydrogen? Explain your answer.

2. Write the balanced symbol equation for the reaction of potassium with water.

3. Write the balanced symbol equation for the reaction between magnesium and lead(II) oxide.

4. Can carbon displace magnesium from magnesium oxide? Explain your answer.

5. The equation for a displacement reaction is shown below:

   $$Mg(s) + Zn^{2+}(aq) \rightarrow Mg^{2+}(aq) + Zn(s)$$

   Are the magnesium atoms oxidised or reduced? Explain your answer.

Developing investigative skills

A student was asked to carry out some possible displacement reactions. The student was given samples of four metals and a dilute solution of each of the metal nitrates. The student set up a series of test tube reactions as summarised in the table below:

Solution	Metal A	Metal B	Metal C	Metal D
Metal A nitrate $A(NO_3)_2$ (aq)		Yes	Yes	No
Metal B nitrate $B(NO_3)_2$ (aq)	No		No	No
Metal C nitrate $C(NO_3)_2$ (aq)	No	Yes		No
Metal D nitrate $D(NO_3)_2$ (aq)	10	11	12	

The student decided that she would need 12 test tubes. In each test tube the student put a 1 cm depth of one of the solutions and then added a small piece of one of the metals. The student left the tubes for 10 minutes and then examined the solution and the piece of metal to see if any reaction was evident. The student then recorded a 'yes' if a displacement reaction had taken place and a 'no' where no reaction was evident. The student didn't have time to record her results for the metal D nitrate solution (tubes 10, 11 and 12).

PROTECTING AGAINST RUST

Rusting is a chemical reaction between iron, water and oxygen. Water and oxygen must both be present for rusting to occur. The process is speeded up if there are also electrolytes such as sodium chloride in the water. This is why rusting takes place much faster in sea water.

The rusting of iron and steel can be prevented in a number of ways. The use of grease, oil or paint are examples of **barrier methods** and prevent water and oxygen from reaching the metal surface. Plastic coating does the same. However, these coatings can be damaged and then rusting will take place.

Iron can be prevented from rusting by using what we know about the reactivity series. Zinc is above iron in the reactivity series; that is, zinc reacts more readily than iron.

Galvanised iron is iron that is coated with a layer of zinc. To begin with, the coating will protect the iron. If the coating is damaged or scratched, the iron is still protected from rusting. This is because zinc is more reactive than iron and so it reacts and corrodes instead of the iron.

If zinc blocks are attached to the hulls of ships, they will corrode instead of the hull. The zinc is called a sacrificial anode and this is an example of **sacrificial protection**.

△ Fig. 2.25 Galvanised iron or steel resists corrosion by air and water.

QUESTIONS

1. What conditions are necessary for iron to rust?

2. What is the disadvantage of using grease to prevent the rusting of iron?

3. a) What is galvanising?

   b) Why is galvanised iron or steel still protected from corrosion when its surface has been scratched?

EXTENSION

The reactivity series of metals is useful to scientists in determining, ways of preventing corrosion. Use the reactivity series to help answer the questions.

1. Iron is a metal in common use but unfortunately it rusts.

   a) Give the conditions needed for iron to rust.

   b) Chromium is sometimes used in chromium plating to prevent iron from rusting. Give a reason for this.

2. Aluminium's position in the reactivity series indicates that it is a highly reactive metal. However, aluminium does not corrode readily. Suggest a reason for this.

Developing investigative skills

A student investigated the reaction of some metals with samples of dilute hydrochloric acid and dilute sulfuric acid. The student added a 3 cm depth of dilute hydrochloric acid to each of four test tubes and added a 3 cm depth of dilute sulfuric acid to another four test tubes. The student then added a 2 cm strip of magnesium ribbon to one test tube of dilute hydrochloric acid and another 2 cm strip to one test tube of dilute sulfuric acid and observed the two reactions. The student then repeated the procedure with two pieces of granulated zinc, two iron nails and two pieces of copper foil. The student recorded the results in the table below.

Metal	Observations with dilute hydrochloric acid	Observations with dilute sulfuric acid
Magnesium ribbon	Rapid effervescence and the ribbon disappeared completely	Rapid effervescence and the ribbon disappeared completely
Zinc (granulated)	Effervescence and the zinc became smaller but did not disappear	Effervescence and the zinc became smaller but did not disappear
Iron nail	Very slow effervescence and little change to the nail	Slow effervescence and little change to the nail
Copper foil	No reaction	No reaction

Make observations and measurements

❶ What does the term 'effervescence' describe?

Analyse and interpret data

❷ What do the results indicate about the reactivity of the two acids? Explain your answer.

❸ What do the results indicate about the reactivity of the four metals? Explain your answer.

Evaluate data and methods

❹ In what ways did the student ensure the comparison between the two acids was fair?

❺ In what way was the comparison between the four metals unfair?

End of topic checklist

A **displacement reaction** is one in which one element takes the place of another element in a compound and removes (displaces) it from the compound.

Galvanising is the process of coating a metal (usually iron) with zinc.

Sacrificial protection involves putting a metal in contact with another more reactive metal so that the more reactive metal oxidises in preference to the less reactive metal.

A **redox** reaction involves both oxidation and reduction.

The facts and ideas that you should know and understand by studying this topic:

○ Know the order of reactivity of metals: potassium, sodium, lithium, calcium, magnesium, aluminium, zinc, iron, copper, silver and gold.

○ Describe how the reactions of metals with water and dilute acids can be used to deduce the order of the reactivity series.

○ Understand how displacement reactions between metals and metal oxides and between metals and metal salts in aqueous solution can be used to deduce the order of reactivity of metals.

○ Understand that oxidation is the addition of oxygen or the loss of electrons.

○ Understand that reduction is the removal of oxygen or the gain of electrons.

○ Understand that an oxidising agent will oxidise another element or compound but will itself be reduced.

○ Understand that a reducing agent will reduce another element or compound but will itself be oxidised.

○ Be able to describe that air (oxygen) and water must be present for iron to rust.

○ Be able to describe how rusting can be prevented by using barrier methods such as grease, oil, plastic or by galvanising.

○ Understand the sacrificial protection of iron in terms of the reactivity series.

End of topic questions

1. Arrange the following metals in order of reactivity, starting with the most reactive first:

   calcium, copper, magnesium, sodium, zinc   (2 marks)

2. This question is about the reaction between magnesium and lead(II) oxide.

   a) Write a balanced symbol equation for the reaction.   (2 marks)

   b) Which is the more reactive metal, magnesium or lead?   (1 mark)

   c) This is an example of a redox reaction. Explain the term 'redox'.   (1 mark)

3. This question is about four metals represented by the letters Q, X, Y and Z. A series of displacement reactions was carried out and the results are shown below:

   Reaction 1: Q oxide + Y → Y oxide + Q

   Reaction 2: X oxide + Z → Z oxide + X

   Reaction 3: Q oxide + Z → no change

   a) Arrange the metals in order of reactivity starting with the most reactive.

   (2 marks)

   b) In reaction 1:

   i) Which substance has been oxidised?   (1 mark)

   ii) Which substance has been reduced?   (1 mark)

   iii) Which substance is the oxidising agent?   (1 mark)

   iv) Which substance is the reducing agent?   (1 mark)

4. Copper(II) sulfate solution reacts with zinc as shown below:

   $CuSO_4(aq) + Zn(s) \rightarrow ZnSO_4(aq) + Cu(s)$

   a) What type of chemical reaction is this?   (1 mark)

   b) What can be deduced about the relative reactivity of copper and zinc?   (1 mark)

5. Zinc prevents the hulls of iron ships from rusting (corroding).

   a) What is the name for this way of preventing rusting?   (1 mark)

   b) How does zinc protect iron from rusting?   (2 marks)

   c) Name three other methods of preventing iron from rusting.   (3 marks)

Extraction and uses of metals

INTRODUCTION

Most metals react easily with other elements and form compounds. Therefore most metals must be extracted from one of their compounds if a pure sample is needed. There are exceptions, and some pure metals can be found in the Earth's crust. For other metals the method of extraction depends on their reactivity. Obtaining iron from iron ore and then converting the iron into steel was a key achievement of the Industrial Revolution. 'Super-metal' aluminium came much later because scientists had to find a different way of extracting it.

△ Fig. 2.26 The manufacture of steel revolutionised construction.

KNOWLEDGE CHECK

✓ Know that metals can be arranged in a reactivity series.
✓ Understand oxidation and reduction in terms of the addition and removal of oxygen.
✓ Understand that ionic compounds conduct electricity when molten or dissolved in water.

LEARNING OBJECTIVES

✓ Know that most metals are extracted from ores found in the Earth's crust and that unreactive metals are often found as the uncombined element.
✓ Be able to explain how the method of extraction of a metal is related to its position in the reactivity series, using as examples the extraction of iron from its oxide using carbon and electrolysis for aluminium.
✓ Be able to comment on a metal extraction process, given appropriate information.
✓ Be able to explain the uses of aluminium, copper, iron and steel in terms of their properties.
✓ Know that an alloy is a mixture of a metal and one or more elements, usually other metals or carbon.
✓ Be able to explain why alloys are harder than pure metals.

HOW ARE METALS EXTRACTED?

Metals are found in the form of **ores** containing minerals mixed with rock. In almost all cases, the mineral is a compound of the metal, not the pure metal. One exception is gold, which exists naturally in a pure state because it is very unreactive.

Extracting a metal from its ore usually involves two steps:

1. The mineral is physically separated from unwanted rock.

2. The mineral is chemically broken down to obtain the metal.

THE EXTRACTION OF METALS

Δ Fig. 2.27 This gold shoulder cape from North Wales is nearly 4000 years old and still in good condition.

The reactivity of a metal determines how it can be extracted from the Earth's crust. It also explains why some metals have been used for thousands of years while others have only been used much more recently.

The most unreactive metals can be found in their native state, that is as the pure metal, and not combined with other elements. Examples of such metals include gold and silver – metals that have been used for thousands of years. It is estimated that gold was first discovered about 5000 years ago.

Metals below carbon in the reactivity series that do not occur naturally can be extracted by heating their ores with carbon. Examples include lead and iron. It is possible that lead was discovered by accident when the silvery element was seen in the ashes of a wood fire that had been made above a deposit of lead ore. It is estimated that lead was first discovered about 4000 years ago.

The most reactive metals, all those above carbon in the reactivity series, have to be extracted from their minerals by electrolysis. This process is a much more recent development and explains why these metals were not used until relatively recently. Aluminium was first extracted in 1825.

REACTIVITY OF METALS

The chemical method chosen to break down a mineral depends on the reactivity of the metal. The more reactive a metal is, the harder it is to break down its compounds. The more reactive metals are obtained from their minerals by electrolysis of a compound in the molten state.

The less reactive metals can be obtained by heating their oxides with carbon. This method only works for metals below carbon in the reactivity series. It involves the reduction of a metal oxide to the metal.

Metal	Extraction method
potassium sodium calcium magnesium aluminium	The most reactive metals are obtained using electrolysis of a compound in the molten state.
(carbon)	
zinc iron tin lead copper	These metals are below carbon in the reactivity series and so can be obtained by heating their oxides with carbon.
silver gold	The least reactive metals are found as pure elements.

△ Table 2.4 Extraction methods of metals.

EXTRACTING ALUMINIUM

Aluminium oxide has a very high melting point and energy costs would be extremely high to melt the oxide in order for the electrolysis to be possible. Dissolving the aluminium oxide in molten cryolite significantly reduces the temperature that is required and so saves on energy costs. In the molten state the Al^{3+} and O^- ions are free to move when an electric current is passed through the mixture.

Aluminium is extracted from the ore bauxite. Aluminium oxide is extracted from bauxite by purification. The anodes are made from carbon and the cathode is the carbon-lined steel case.

At the cathode aluminium is formed:

aluminium ions	+	electrons	→	aluminium
$Al^{3+}(l)$	+	$3e^-$	→	$Al(l)$

The aluminium ions are reduced by the addition of electrons to form aluminium atoms.

At the anode oxygen is formed:

oxide ions	→	oxygen molecules	+	electrons
$2O^{2-}(l)$	→	$O_2(g)$	+	$4e^-$

The oxide ions are oxidised as they lose electrons to form oxygen molecules. The oxygen reacts with the carbon anodes to form carbon dioxide. Because of this, the rods need to be replaced constantly. The process uses a great deal of electricity and is not cost-efficient unless the electricity is cheap. Aluminium is often extracted in countries with well-developed hydroelectric power.

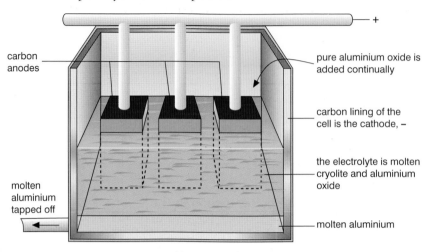

carbon anodes

pure aluminium oxide is added continually

carbon lining of the cell is the cathode, –

the electrolyte is molten cryolite and aluminium oxide

molten aluminium tapped off

molten aluminium

△ Fig. 2.28 Extracting aluminium is expensive. Molten cryolite is used to produce an electrolyte that has a lower melting point than that of pure aluminium oxide.

THE USES OF ALUMINIUM

The uses of aluminium are linked directly to its properties.

Uses of aluminium	Properties making aluminium suitable for the use
Packaging – drinks cans, foil wrapping, foil containers	Non-toxic Impermeable – no aroma or loss of flavour Resistant to corrosion
Transport – aeroplanes	High strength to weight ratio Low density Resistant to corrosion Note: Alloys are often used because they are stronger than pure aluminium.
Overhead electrical cables	High electrical conductivity Low density
As a building material	Easily shaped Low corrosion High strength to weight ratio
For kitchen utensils	Shiny appearance Non-corrosive Non-toxic

△ Table 2.5 Uses of aluminium.

QUESTIONS

1. Name a metal that is found as a pure element.

2. Name a metal other than aluminium that is extracted from one of its minerals by electrolysis of a molten compound.

3. Why is aluminium not extracted by heating aluminium oxide with carbon?

4. Read the information about the extraction of aluminium using electrolysis on page 145.

   a) Why is cryolite used?

   b) At which electrode do the aluminium ions form aluminium atoms?

   c) Write a half-equation for the formation of aluminium atoms from aluminium ions.

   d) Are the aluminium ions oxidised or reduced as they form aluminium atoms? Explain your answer.

   e) Why is it necessary to regularly replace the carbon electrodes?

5. What properties of aluminium make it a suitable material for constructing an aeroplane?

SCIENCE IN CONTEXT

FACTS ABOUT ALUMINIUM

1. Aluminium is the most abundant metal in the Earth's crust, and the third most abundant element overall, after oxygen and silicon. It makes up about 8% by mass of the Earth's solid surface. Aluminium metal is too reactive to occur in nature. Instead, it is found combined in over 270 different minerals. The main ore of aluminium is bauxite. Bauxite is mined extensively to meet the demand for aluminium: Australia produced 62 million tonnes of bauxite in 2005.

2. Impurities in aluminium oxide, such as chromium or iron, yield the gemstones ruby and sapphire.

3. The cost of electricity represents about 20% to 40% of the total cost of producing aluminium, depending on the location of the smelter. Smelters tend to be situated where electric power is both plentiful and inexpensive, such as in the United Arab

△ Fig. 2.29 Worldwide we use 6 billion aluminium cans each year, using about 200 000 tonnes of aluminium.

Emirates where there are excess natural gas supplies, and Iceland and Norway with energy generated from renewable sources such as hydroelectric power. Aluminium production consumes roughly 5% of the electricity generated in the USA.

4. The corrosion resistance of aluminium is due to a thin surface layer of aluminium oxide that forms when the metal is exposed to air, effectively preventing further oxidation.

5. Aluminium is 100% recyclable without any loss of its natural qualities. Recycling involves melting the scrap, which requires only 5% of the energy used to produce aluminium from its ore, although a significant part (up to 15% of the input material) is lost as dross (an ash-like oxide). However, the dross can undergo a further process to extract more aluminium.

EXTRACTING IRON

Iron is produced on a very large scale by reduction using carbon. The reaction takes place in a huge furnace called a blast furnace. (You don't need to know the details of this.)

Three important raw materials are put in the top of the furnace: iron ore (iron(III) oxide, the source of iron), coke (the source of carbon needed for the reduction) and limestone, needed to remove the impurities as slag. Iron ore is also known as haematite.

△ Fig. 2.30 Iron ore (haematite).

△ Fig. 2.31 Coke.

△ Fig. 2.32 Limestone.

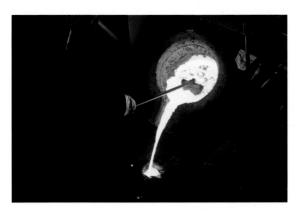

△ Fig. 2.33 Molten iron.

△ Fig. 2.34 Slag.

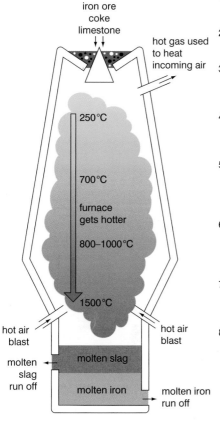

iron ore
coke
limestone

hot gas used
to heat
incoming air

250°C

700°C

furnace
gets hotter

800–1000°C

1500°C

hot air
blast

hot air
blast

molten
slag
run off

molten slag

molten iron

molten iron
run off

1 Iron ore, coke and limestone are fed into the top of the blast furnace

2 Hot air is blasted up the furnace from the bottom

3 Oxygen from the air reacts with coke to form carbon dioxide:
$$C(s) + O_2(g) \longrightarrow CO_2(g)$$

4 Carbon dioxide reacts with more coke to form carbon monoxide:
$$CO_2(g) + C(s) \longrightarrow 2CO(g)$$

5 Carbon monoxide is a reducing agent. Iron(III) oxide is reduced to iron:
┌─ reduction = loss of oxygen ─┐
$$Fe_2O_3(s) + 3CO(g) \longrightarrow 2Fe(l) + 3CO_2(g)$$

6 Dense molten iron runs to the bottom of the furnace and is run off. There are many impurities in iron ore. The limestone helps to remove these as shown in 7 and 8.

7 Limestone is broken down by heat to calcium oxide:
$$CaCO_3(s) \longrightarrow CaO(s) + CO_2(g)$$

8 Calcium oxide reacts with impurities like sand (silicon dioxide) to form a liquid called 'slag':
$$CaO(s) + SiO_2(s) \longrightarrow CaSiO_3(l)$$
impurity   slag
The liquid slag falls to the bottom of the furnace and is tapped off.

△ Fig. 2.35 How iron is extracted in a blast furnace.

The overall reaction is:

iron(III) oxide	+	carbon	→	iron	+	carbon dioxide
$2Fe_2O_3(s)$	+	$3C(s)$	→	$4Fe(l)$	+	$3CO_2(g)$

The reduction happens in three stages.

Stage 1: The coke (carbon) reacts with oxygen 'blasted' into the furnace.

carbon	+	oxygen	$\rightarrow$	carbon dioxide
$C(s)$	+	$O_2(g)$	$\rightarrow$	$CO_2(g)$

Stage 2: The carbon dioxide is reduced by unreacted coke to form carbon monoxide:

carbon dioxide	+	carbon	$\rightarrow$	carbon monoxide
$CO_2(g)$	+	$C(s)$	$\rightarrow$	$2CO(g)$

Stage 3: The iron(III) oxide is reduced by the carbon monoxide to iron, and the carbon monoxide is oxidised to carbon dioxide:

iron(III) oxide	+	carbon monoxide	$\rightarrow$	iron	+	carbon dioxide
$Fe_2O_3(s)$	+	$3CO(g)$	$\rightarrow$	$2Fe(s)$	+	$3CO_2(g)$

EXTENSION

1. The thermit process can be used to extract iron from iron(III) oxide using aluminium metal.

   a) Write a balanced symbol equation for the extraction of iron using this process.

   b) Using the reactivity series, explain why this reaction occurs.

   c) Name two other metals that could be used to extract iron from its oxide indicating whether the reactions would be more or less reactive than using aluminium.

   d) This reaction could be described as either a displacement or a redox reaction. Explain why.

MAKING STEEL FROM IRON

Iron from the blast furnace is brittle because it contains a large percentage (usually 4%) of carbon (from the coke). It also rusts very easily. Because of this, most iron is converted into steel, an alloy of iron.

In steel making, the molten iron straight from the blast furnace is heated and oxygen is passed through it to remove some of the large percentage of carbon present after the iron is extracted:

$C(s) + O_2(g) \rightarrow CO_2(g)$

Steel is iron with 0.1–1.5% carbon content. Steel is more resistant to corrosion and is less brittle than iron. It has a wide range of uses, depending on its carbon content. For example:

- low carbon (<0.3%): car bodies
- medium carbon (0.3–0.9%): railway tracks
- high carbon (0.9–1.5%): knives

Stainless steels are made by adding a wide range of metals to steel such as chromium, nickel, vanadium and cobalt. Each one gives the steel particular properties for specific uses. For example, vanadium steel is used to make high-precision, hard-wearing industrial tools.

Alloys

An **alloy** is a mixture of a metal with one or more other elements.

The reason for producing alloys is to 'improve' the properties of a metal. Table 2.6 shows some examples.

Alloy	Property improved
Steel	Hardness/tensile strength
Bronze	Hardness
Solder	Lower melting point
Cupronickel	Shiny and hardwearing (used for coins)
Stainless steel	Resistance to corrosion
Brass	Easier to shape and stamp into shape

△ Table 2.6 Alloys and their properties.

△ Fig. 2.36 Alloys are used to make coins.

The structure of alloys

The structure of pure metallic elements is usually shown as in Fig. 2.36.

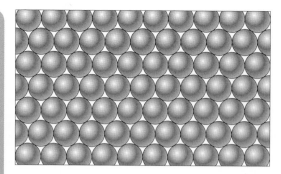

△ Fig. 2.37 Particles in a solid.

This is a simplified picture but, surprisingly, such a structure is very weak. If there is the slightest difference between the planes of atoms, the metal will break at that point.

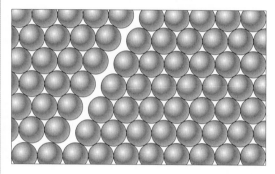

△ Fig. 2.38 The gaps show a weak point of a metal.

The more irregular (jumbled-up) the metal atoms are, the stronger the metal is. The irregular atom arrangement prevents the rows of metal atoms from readily sliding over each other – the added elements break up the regular rows of metal atoms, as can be seen in Fig. 2.39 and Fig. 2.40. This is why alloy structures are stronger: because of the elements added.

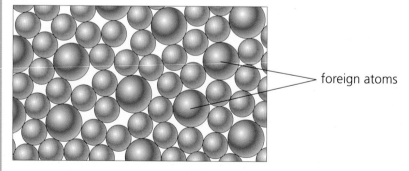

foreign atoms

△ Fig. 2.39 Atoms in an alloy.

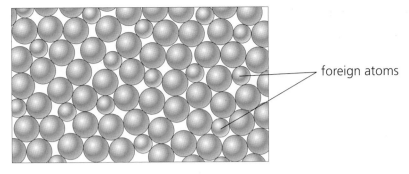

foreign atoms

△ Fig. 2.40 Even smaller atoms make the metal stronger.

Steel that is heated to red heat and then plunged into cold water is made harder by the process of 'jumbling up' the metal atoms. Further heat treatment is used to increase the strength and toughness of the alloy.

QUESTIONS

1. Iron is produced in a blast furnace by reduction of its ore. The most common ore of iron is iron(III) oxide, Fe_2O_3. The furnace is heated by burning coke (an impure form of carbon) in air.

   a) Give a reason why iron can be obtained from its ore by heating with coke.

   b) Write a balanced symbol equation to show the reduction of iron(III) oxide by carbon.

2. This question is about steel.

   a) Steel is an alloy. What is an alloy?

   b) The iron produced in the blast furnace contains a relatively high proportion of carbon. How is this proportion of carbon reduced when iron is converted into steel?

   c) Why is steel often used instead of the iron produced in the blast furnace?

Uses of copper

Copper is an excellent conductor of electricity and this is why it is used in electrical cables.

Copper is also an excellent conductor of heat with a high melting point. These properties make it ideal for cooking utensils.

EXTENSION

The method used to extract a metal from its ores depends on the reactivity of the metal. The more reactive the metal the harder and possibly more expensive it is to extract from its ore. Titanium is corrosion resistant, very strong and has a relatively low density, but because its extraction is very expensive it is only used for specialised purposes. Use your knowledge and the information given to answer the questions.

1. Metals such as iron, aluminium and titanium need to be extracted from their ores. Ores are commonly oxides, rutile TiO_2 is the commonly occurring ore of titanium.

a) Give the names of the common ores of iron and aluminium and the formulae of the oxides.

b) Titanium oxide can not be reduced by using carbon. Give a reason for this.

c) Titanium is expensive because it is awkward to extract. The steps given below outline the extraction of titanium:

Step 1: Rutile needs to be heated with chlorine and coke to form titanium(IV) chloride.

Step 2: In the UK titanium(IV) chloride is reduced to titanium metal using sodium, whereas in other areas of the world magnesium is used.

i) Write a word and balanced symbol equation for step 1.

ii) Write a word and balanced symbol equation for step 2.

iii) Name the type of reaction that occurs in step 2.

iv) Explain why sodium metal is used in this reaction.

v) Suggest what conditions would be necessary if sodium is used rather than magnesium and why it is easier to use magnesium.

vi) Copper is a metal that is readily available. Suggest why this is not used to extract titanium from its ore.

End of topic checklist

An **alloy** is a mixture of a metal and one or more other elements.

Electrolysis is the breaking down of a compound by passing an electric current through it.

An **electrolyte** is a substance that allows the passage of an electric current when molten or dissolved in water.

The facts and ideas that you should know and understand by studying this topic:

○ Know that most metals are extracted from ores and that unreactive metals are often found in the Earth's crust as the uncombined element, not as an ore.

○ Be able to explain how the methods of extraction of metals relate to their positions in the reactivity series.

○ Be able to comment on a method of metal extraction, when given information about the process.

○ Be able to describe and explain the extraction of aluminium from purified aluminium oxide by electrolysis.

○ Be able to write the half-equations for the reactions of aluminium and oxide ions at the cathode and anode (electrodes).

○ Be able to write balanced chemical equations for reactions involved in the extraction of iron from iron ore, when given appropriate information.

○ Know the uses of aluminium and be able to explain these in terms of its properties.

○ Know the uses of iron and be able to explain these in terms of its properties.

○ Know the uses of copper and explain in terms of its properties.

○ Know that an alloy is a mixture of a metal and one or more elements, usually other metals or carbon.

○ Be able to explain why alloys are harder than pure metals.

○ Know the uses of low-carbon, high-carbon and stainless steel.

End of topic questions

1. The least reactive metals such as gold and silver are found in their native state. What do you understand by this? **(1 mark)**

2. Iron is extracted from iron ore (iron(III) oxide) in a blast furnace by heating with coke (carbon).

    a) Write a balanced symbol equation including state symbols for the overall reaction. **(2 marks)**

    b) Is the iron(III) oxide oxidised or reduced in this reaction? Explain your answer. **(1 mark)**

3. Zinc can also be extracted from zinc oxide by heating with carbon.

    a) Write a balanced symbol equation, including state symbols, for this reaction. **(2 marks)**

    b) Zinc could also be extracted by the electrolysis of molten zinc oxide. Suggest why heating with carbon is the preferred method of extraction. **(2 marks)**

4. Aluminium is extracted from aluminium oxide (Al_2O_3) by electrolysis. Aluminium oxide contains Al^{3+} and O^{2-} ions. The aluminium oxide is dissolved in molten cryolite.

    a) Why is the electrolysis carried out in a solution of aluminium oxide rather than solid aluminium oxide? **(2 marks)**

    b) Write a half-equation to show how O^{2-} ions are converted into oxygen gas. **(2 marks)**

    c) Write a half-equation to show the formation of aluminium at the cathode. **(2 marks)**

    d) The extraction of aluminium often takes place in areas with easy access to hydroelectric power. Suggest a reason for this. **(2 marks)**

5. Explain what is meant by the following terms:

    a) electrolysis **(1 mark)**

    b) electrolyte. **(1 mark)**

6. Sodium can be extracted by the electrolysis of molten sodium chloride. Sodium chloride has a giant ionic lattice structure made up of sodium ions, Na^+, and chloride ions, Cl^-.

   a) What is a giant ionic lattice structure? **(2 marks)**

   b) Explain why the sodium chloride has to be in a molten state for electrolysis to take place. **(1 mark)**

   c) At which electrode will the sodium ions be discharged? **(1 mark)**

   d) Write a half-equation to show the discharge of the sodium ions. **(1 mark)**

   e) Which electrode will the chloride ions be discharged at? **(1 mark)**

   f) Write a half-equation to show the discharge of the chloride ions. **(2 marks)**

7. Steel is an alloy of iron.

   a) What is an alloy? **(1 mark)**

   b) What is the composition of steel? **(1 mark)**

8. Suggest reasons for the following:

   a) Aluminium is often used instead of copper in overhead electrical cables. **(2 marks)**

   b) Aluminium foil is used in food packaging. **(2 marks)**

   c) Aluminium is used in aircraft construction. **(2 marks)**

   d) Steel rather than iron is used in building construction. **(2 marks)**

   e) Copper is a suitable metal for the following uses:

      i) As electrical wiring;

      ii) As the base of a saucepan. **(2 marks)**

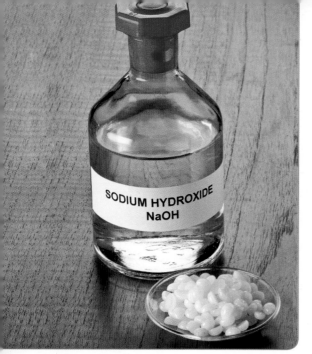

△ Fig. 2.41 Sodium hydroxide

Acids, alkalis and titrations

INTRODUCTION

Acids are commonly used in everyday life. Concentrated hydrochloric acid is corrosive to skin and eyes and an irritant to the respiratory system. It may produce toxic gases (chlorine and hydrogen chloride gas) during procedures or reactions. Concentrated sulfuric acid is corrosive to skin and eyes. About 20 million tonnes of hydrochloric acid are manufactured worldwide each year. Alkalis and bases are less common in everyday use, yet about 60 million tonnes of sodium hydroxide are produced worldwide each year and used in the manufacture of paper and soap. Sodium hydroxide is corrosive to skin and eyes, and by inhalation.

KNOWLEDGE CHECK

✓ Know the names of some common acids, including hydrochloric and sulfuric acids.
✓ Know that vegetable dyes can be used as indicators to identify acids and alkalis.
✓ Be able to use state symbols such as (s), (l), (g) and (aq).

LEARNING OBJECTIVES

✓ Describe the use of litmus, phenolphthalein and methyl orange to distinguish between acidic and alkaline solutions.
✓ Understand how the 0–14 pH scale can be used to identify acids and alkalis and their strengths.
✓ Be able to describe the use of universal indicator to measure the approximate pH value of an aqueous solution.
✓ Know that acids in an aqueous solution are a source of hydrogen ions and alkalis in an aqueous solution are a source of hydroxide ions.
✓ Know that alkalis can neutralise acids.
✓ Be able to describe how to carry out an acid–alkali titration.

AQUEOUS SOLUTIONS

When any substance dissolves in water, it forms an aqueous solution, shown by the state symbol (aq). Aqueous solutions can be acidic, alkaline or neutral.

Indicators are used to tell if a solution is acidic, alkaline or neutral. They can be used either as liquids or in paper form, and they turn different colours with different solutions. There are different indicators that can be used.

The most common indicator is litmus. Its colours are shown in the table, together with those for phenolphthalein and methyl orange indicators.

Indicator	Colour in acidic solution	Colour in alkaline solution
litmus	red	blue
phenolphthalein	colourless	pink
methyl orange	pink	yellow

△ Table 2.7 Indicators.

Universal indicators can show how strongly acidic or how strongly alkaline a solution is because they have more colours than litmus. Each colour is linked to a number ranging from 0 (most strongly acidic solution) to 14 (most strongly alkaline solution). This range is called the **pH scale**.

Concentration of hydrogen ions compared to distilled water		Examples of solutions and their respective pH
1/10 000 000	14	Liquid drain cleaner, caustic soda
1/1 000 000	13	Bleaches oven cleaner
1/100 000	12	Soapy water
1/10 000	11	Household ammonia (11.9)
1/1000	10	Milk of magnesium (10.5)
1/100	9	Toothpaste (9.9)
1/10	8	Baking soda (8.4), seawater, eggs
1	7	"Pure" water (7)
10	6	Urine (6) milk (6.6)
100	5	Acid rain (5.6) black coffee (5)
1000	4	Tomato juice (4.1)
10000	3	Grapefruit and orange juice, soft drink
100000	2	Lemon juice (2.3) vinegar (2.9)
1 000 000	1	Hydrochloric acid secreted from the stomach lining (1)
10 000 000	0	Battery acid

A neutral solution has a pH of 7

△ Fig. 2.42 The pH scale.

WHAT ARE ACIDS?

Acids are substances that contain replaceable hydrogen atoms. These hydrogen atoms are replaced in chemical reactions by metal atoms. Acids have pHs in the range 0–6.

Acids only show their acidic properties when water is present. This is because, in water, acids form hydrogen ions, $H^+(aq)$

(which are also protons), and it is these ions that create acidic properties. For example:

$$HCl(aq) \rightarrow H^+(aq) + Cl^-(aq)$$

WHAT ARE BASES AND ALKALIS?

The oxides and hydroxides of metals are called **bases**.

If the oxide or hydroxide of a metal dissolves in water, it is also called an **alkali**. For example:

sodium	+	oxygen	→	sodium oxide
$4Na(s)$	+	$O_2(g)$	→	$2Na_2O(s)$

sodium oxide	+	water	→	sodium hydroxide
$Na_2O(s)$	+	$H_2O(l)$	→	$2NaOH(aq)$

The sodium oxide above is a base because it is the oxide of the metal sodium. In addition, it dissolves in water to form the alkali sodium hydroxide. A weakly alkaline solution has a pH between 8 and about 10. A strongly alkaline solution has a pH between about 10 and 14.

Alkalis are substances that dissolve in water to form hydroxide ions, $OH^-(aq)$. For example:

$$NaOH(aq) \rightarrow Na^+(aq) + OH^-(aq)$$

Another common alkali is potassium hydroxide, KOH:

$$KOH(aq) \rightarrow K^+(aq) + OH^-(aq)$$

QUESTIONS

1. Two solutions are tested with universal indicator paper. Solution A has a pH of 8 and solution B has a pH of 14. What does this tell you about the two solutions?

2. Methyl orange is added to a solution and the solution turns pink. What does this tell you about the solution?

3. Calcium oxide is an example of a base. How do you know this?

SCIENCE IN CONTEXT

SOME INTERESTING FACTS ABOUT ACIDS AND ALKALIS

1. Pure sulfuric acid is a clear, oily, highly corrosive liquid. It was well known to Islamic, Greek and Roman scholars in the ancient world, when it was called 'oil of vitriol'. Although pure sulfuric acid does not occur naturally on Earth because of its attraction to water, dilute sulfuric acid is found in acid rain and in the upper atmosphere of the planet Venus. It has a wide range of industrial uses, from making fertilisers, dyes, paper and pharmaceuticals to batteries, steel and iron. Sulfuric acid is not toxic, but it is highly reactive with water, producing a strong exothermic reaction. It can cause severe burns. The acid must be stored in glass or plastic containers (never metal) and handled with extreme care.

2. Hydrochloric acid, although classified as corrosive to skin and eyes, and irritating to the respiratory system, is part of the gastric acid in the stomach involved in the digestive process. Excess acid in the stomach can cause indigestion but 'anti-acid' (alkali) medications can be taken to neutralise this.

3. Perhaps the strongest acid is a mixture of nitric acid and hydrochloric acid, known as 'aqua regia' because it reacts with the 'royal' metals. Unlike other acids it will react with very unreactive metals such as gold and platinum. However, some metals such as titanium and silver are not affected.

4. Formic acid (now called methanoic acid) is in the venom of ant and bee stings. Such stings can be relieved by neutralising with an alkali such as sodium bicarbonate (sodium hydrogen carbonate). Wasp stings, however, contain an alkali, and so need to be neutralised by a weak acid such as vinegar.

The differences can be remembered using:

Bee - **B**icarb

Wasp (W looks like two vs) – **V**inegar

The difficulty is adding the right amount to exactly neutralise the sting – otherwise you simply replace one irritant with another!

△ Fig. 2.43 A wasp.

5. Sodium hydroxide, although well known as corrosive to the skin and eyes and by inhalation, is used to manufacture soap and as a drain cleaner. Surprisingly, it used to be used as a 'relaxer' to straighten hair. The high incidence of chemical burns soon put an end to this practice!

Neutralisation describes the reactions of acids with alkalis and bases. When acids react with alkalis, the reaction is between H⁺ ions and OH⁻ ions to make water, as in:

$$H^+(aq) + OH^-(aq) \rightarrow H_2O(l)$$

Reactions of acids with alkalis are used in the experimental procedure of **titration**, in which solutions react together to give the end-point shown by an indicator. Calculations are then performed to find the **concentration** of the acid or the alkali.

Acid–alkali titration

Titration is an accurate experimental method of finding the concentration of an acid or an alkali using the neutralisation reaction.

Method:

1. 25.0 cm³ of the alkali is put into a conical flask using a pipette. The flask is placed on a white tile (to see colour changes more easily) and a few drops of an appropriate indicator (methyl orange) are added.

2. The burette is filled with an acid.

3. When the burette tap opens, the acid runs into the alkali in the flask and the flask is swirled.

4. Near the 'end-point' (neutralisation, when the indicator changes colour), the solution is added drop by drop until one drop changes the colour of the indicator from yellow to pink.

5. The volume of solution added from the burette is read and the whole process repeated until burette readings that are identical (or within 0.10 cm³) are obtained.

6. The burette volume is used to calculate the concentration of the alkali/acid.

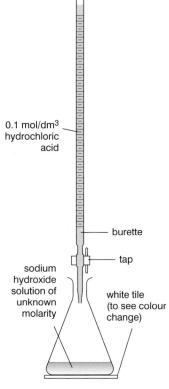

0.1 mol/dm³ hydrochloric acid

burette

tap

sodium hydroxide solution of unknown molarity

white tile (to see colour change)

△ Fig. 2.44 Titration apparatus.

WORKED EXAMPLE

25.00 cm³ of sodium hydroxide solution required 30.00 cm³ of 0.2 mol/dm³ hydrochloric acid for neutralisation.

What is the concentration of the sodium hydroxide solution?

$$HCl(aq) + NaOH(aq) \rightarrow NaCl(aq) + H_2O(l)$$

Moles HCl added = $\dfrac{30.00}{1000} \times 0.2 = 0.0060$

From the equation the ratio of HCl : NaOH is 1:1 so

Moles NaOH = 0.0060

Concentration of NaOH = $\dfrac{1000 \times 0.006}{25.00} = 0.24$ mol/dm³

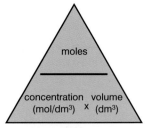

moles

concentration   volume
(mol/dm³)   ×   (dm³)

△ Fig. 2.45 This triangle will help you to calculate concentrations of solutions.

QUESTIONS

1. What does the word 'neutralisation' mean?

2. Which ion do all acids form in solution?

3. Which ion do all alkalis form in solution?

Developing investigative skills

A student wanted to find the concentration of a solution of potassium hydroxide. The student decided to use a titration method using a 0.10 mol/dm^3 solution of sulfuric acid. (Note: dilute sulfuric acid at this concentration is not currently classified as hazardous. Between 0.1 mol/dm^3 and 0.4 mol/dm^3 potassium hydroxide is classified as irritating to skin and eyes. Methyl orange indicator is most likely to be classified as highly flammable and harmful due to the solvent that it is made up with, (usually IDA).)

$2KOH(aq) + H_2SO_4(aq) \rightarrow K_2SO_4(aq) + 2H_2O(l)$

The method the student used is listed below:

❶ The student washed a pipette and a burette carefully, making sure they were drained after washing.

❷ The student used the pipette to transfer 25.00 cm^3 of the potassium hydroxide solution into a clean conical flask and added 3 drops of methyl orange indicator.

❸ The student filled the burette with the sulfuric acid solution, making sure that there were no air bubbles in the jet of the burette. Finally the student took the first reading of the volume of acid in the burette.

❹ The student ran the acid into the alkali in the conical flask, swirling the flask all the time. When the student thought the indicator colour was close to changing, the student added the acid more slowly until the colour changed. The student then took the second reading on the burette.

❺ The student repeated the whole procedure twice, making sure that between each experiment the student washed the conical flask carefully. The results are shown in the table.

Volume of potassium hydroxide solution = 25.00 cm^3

Burette reading	1st experiment	2nd experiment	3rd experiment
2nd reading (cm^3)	17.50	19.50	20.50
1st reading (cm^3)	0.00	2.50	3.50
Difference (cm^3)	17.50	17.00	17.00

△ Table 2.8 Results of experiment.

Demonstrate and describe techniques

❶ Describe precisely how the student should have washed a) the pipette and b) the burette, at the beginning of the experiment.

❷ How could the pipette be used most safely to measure out the 25.00 cm³ of potassium hydroxide solution?

❸ How can a burette be filled most easily without spilling acid?

❹ What could the student have done to make the colour change easier to observe? What colour change was the student looking for?

❺ How should the student have washed out the conical flask between experiments to prevent contamination?

Analyse and interpret data

❻ Why are the second and third experiments likely to be more accurate than the first?

❼ What volume of sulfuric acid would you use in your calculation? Explain your answer.

❽ Use the results and the chemical equation to work out the concentration of the potassium hydroxide solution. (Hint: start by writing down the number of moles of H_2SO_4 used.)

End of topic checklist

An **acid** is a substance that contains replaceable hydrogen atoms which form H^+ ions in water.

A **base** is the oxide or hydroxide of a metal.

An **alkali** is a base that is soluble in water and produces OH^- ions.

A **neutralisation reaction** is one in which an acid reacts with a base or alkali to form a salt and water.

Titration is an accurate method for calculating the concentration of an acid or alkali solution in a neutralisation reaction.

The facts and ideas that you should know and understand by studying this topic:

○ Be able to describe how indicators such as litmus, phenolphthalein and methyl orange can be used to distinguish between acidic and alkaline solutions.

○ Know the colours of each of the above indicators in acidic and alkaline solutions.

○ Be able to describe how universal indicator can be used to test a solution.

○ Understand how the pH scale from 0 to 14 can be used to classify solutions as strongly acidic (0–3), weakly acidic (4–6), neutral (7), weakly alkaline (8–10) or strongly alkaline (11–14).

○ Know that acids when in aqueous solution are a source of H^+ ions.

○ Know that alkalis when in aqueous solution are a source of OH^- ions.

○ Understand how to calculate the concentration of an acid or alkali from the results of a titration experiment.

End of topic questions

1. a) What is an indicator? (1 mark)

   b) What is the pH scale? (1 mark)

   c) What do the following pH numbers indicate about a solution that has been tested?

   i) pH 6 (1 mark)

   ii) pH 8 (1 mark)

   iii) pH 14. (1 mark)

2. a) What is an acid? (1 mark)

   b) What is an alkali? (1 mark)

   c) What is the name of the process when an acid reacts with an alkali to form water? (1 mark)

3. Explain how you carry out an acid–alkali titration.

   a) What equipment would you use to measure out 25 cm^3 of the alkali? (1 mark)

   b) What indicator would you use and what colour change would you expect to see at the end-point? (3 marks)

   c) Why is the acid added to the alkali using a burette rather than a measuring cylinder? (1 mark)

   d) In a titration experiment 25 cm^3 of 0.1 mol/dm^3 sodium hydroxide solution reacted with 20.0 cm^3 of hydrochloric acid. Calculate the concentration of the hydrochloric acid used. (7 marks)

Acids, bases and salt preparations

△ Fig. 2.46 These are crystals of sodium chloride which we often refer to as common salt. In fact, sodium chloride is an example of a group of compounds called salts.

INTRODUCTION

This topic builds on the ideas introduced in the previous topic. It reinforces some of the key ideas about acids, alkalis and bases and then concentrates on the reactions that can be used to make a group of compounds called salts. Common salt, sodium chloride, is an example of a salt.

KNOWLEDGE CHECK

✓ Know that acids are a source of H^+ ions.
✓ Know that a base is the oxide or hydroxide of a metal.
✓ Know that an alkali is a source of OH^- ions.
✓ Know that a neutralisation reaction is the reaction between an acid and an alkali.

LEARNING OBJECTIVES

✓ Know the general rules for predicting the solubility of ionic compounds in water:
  • Common sodium, potassium and ammonium compounds are soluble.
  • All nitrates are soluble.
  • Common chlorides are soluble, except those of silver and lead(II).
  • Common sulfates are soluble, except for those of barium, calcium and lead(II).
  • Common carbonates are insoluble, except for those of sodium, potassium and ammonium.
  • Common hydroxides are insoluble except for those of sodium, potassium and calcium (calcium hydroxide is slightly soluble).
✓ Understand acids and bases in terms of proton transfer: an acid is a proton donor and a base is a proton acceptor.
✓ Describe the reactions of hydrochloric acid, sulfuric acid and nitric acid with metals, bases and metal carbonates (excluding the reactions between nitric acid and metals) to form salts.
✓ Know that metal oxides, metal hydroxides and ammonia can act as bases, and that alkalis are bases that are soluble in water.
✓ Describe an experiment to prepare a pure, dry sample of a soluble salt, starting from an insoluble reactant.
✓ Describe an experiment to prepare a pure, dry sample of a soluble salt, starting from an acid and alkali.
✓ Describe an experiment to prepare a pure, dry sample of an insoluble salt, starting from two soluble reactants.

✓ Be able to prepare a sample of pure, dry hydrated copper(II) sulfate crystals starting from copper(II) oxide.

✓ Be able to prepare a sample of pure, dry lead(II) sulfate.

WHAT ARE SALTS?

You saw in the previous topic that acids contain replaceable hydrogen atoms. When metal atoms take their place, a compound called a **salt** is formed. The names of salts have two parts, as shown:

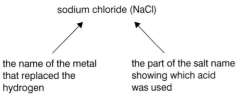

sodium chloride (NaCl)

the name of the metal that replaced the hydrogen

the part of the salt name showing which acid was used

△ Fig. 2.47 A salt – sodium chloride.

Table 2.9 shows the four most common acids and their salt names.

Acid	Salt name
hydrochloric (HCl)	chloride (Cl⁻)
nitric (HNO_3)	nitrate (NO_3^-)
sulfuric (H_2SO_4)	sulfate (SO_4^{2-})
phosphoric (H_3PO_4)	phosphate (PO_4^{3-})

△ Table 2.9 Common acids and their salt names.

Salts are ionic compounds. The names of these compounds are created by taking the first part of the name from the metal ion, which is a positive ion (cation), and the second part of the name from the acid, which is a negative ion (anion). For example:

copper(II) sulfate: Cu^{2+} and $SO_4^{2-} \rightarrow CuSO_4$

Salts are often found in the form of crystals. Salt crystals often contain water of crystallisation, which is responsible for their crystal shapes. Water of crystallisation is shown in the chemical formula of a salt. For example:

copper(II) sulfate crystals: $CuSO_4 . 5H_2O$

iron(II) sulfate crystals: $FeSO_4 . 7H_2O$

△ Fig. 2.48 Copper(II) sulfate crystals.

MAKING SALTS

There are five common methods for making salts. Four of these make soluble salts and one makes insoluble salts.

The solubility of salts in water

Here are the general rules that describe the solubility of common types of salts in water:

- All common sodium, potassium and ammonium salts are soluble.
- All nitrates are soluble.
- Common chlorides are soluble, except silver chloride and lead(II) chloride.
- Common sulfates are soluble, except those of barium, lead(II) and calcium.
- Common carbonates are insoluble, except those of sodium, potassium and ammonium.
- Common hydroxides are insoluble except for those of sodium, potassium and calcium (calcium hydroxide is slightly soluble).

Making soluble salts

1.   acid   + alkali   → a salt   + water

for example: $HCl(aq)$ + $NaOH(aq)$ → $NaCl(aq)$ + $H_2O(l)$

---

2.   acid   + base   → a salt   + water

for example: $H_2SO_4(aq)$ + $CuO(s)$ → $CuSO_4(aq)$ + $H_2O(l)$

---

3.   acid   + carbonate → a salt   + water + carbon dioxide

for example: $2HNO_3(aq)$ + $CuCO_3(s)$ → $Cu(NO_3)_2(aq)$ + $H_2O(l)$ + $CO_2(g)$

---

4.   acid   + metal   → a salt   + hydrogen

for example: $2HCl(aq)$ + $Mg(s)$ → $MgCl_2(aq)$ + $H_2(g)$

Here is a shortcut for remembering the four general equations above. Remember the initials of the reactants:

A (acid) + A (alkali)

A (acid) + B (base)

A (acid) + C (carbonate)

A (acid) + M (metal)

The symbol '(aq)' after the formula of the salt shows that it is a soluble salt.

In the laboratory

Of the four methods for making soluble salts, symbol (aq), only one uses two solutions.

Method 1 (neutralisation):

acid(aq) + alkali(aq) → a salt(aq) + water(l)

The other three methods involve adding a solid(s) to a solution(aq).

Method 2:

acid(aq) + base(s) → a salt(aq) + water(l)

Method 3:

acid(aq) + carbonate(s) → a salt(aq) + water(l) + carbon dioxide(g)

Method 4:

acid(aq) + metal(s) → a salt(aq) + hydrogen(g)

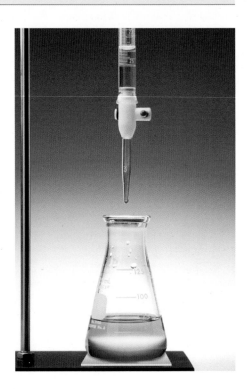

▷ Fig. 2.49 Using the neutralisation method for a titration.

You should know that as acids in water form H$^+$ (aq) ions and alkalis form OH$^-$(aq) ions, the neutralisation reaction of acids with alkalis can be written as:

$$H^+(aq) + OH^-(aq) \rightarrow H_2O(l)$$

Remember that all neutralisation reactions can be represented by this equation, whatever the acid or alkali used.

A common way of defining an acid is as a **proton donor**. An alkali is therefore defined as a **proton acceptor**.

The flow diagram shows how to make soluble salts from solids.

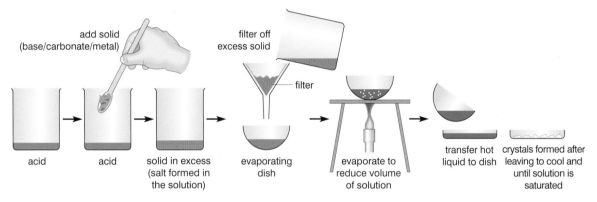

△ Fig. 2.50 Making soluble salts from solids.

QUESTIONS

1. What is a salt?

2. Which acid would you use to make a sample of sodium sulfate?

3. Would you expect potassium chloride to be soluble or insoluble in water? Explain your answer.

4. What is the name of the salt formed when calcium carbonate reacts with nitric acid?

Making insoluble salts

If two solutions of soluble salts are mixed together to form two new salts, and one of the products is insoluble, the insoluble salt forms a precipitate – a solid made in solution. This process is called **precipitation**. The general equation is:

soluble salt + soluble salt → insoluble salt + soluble salt

precipitate

For example:

$Na_2CO_3(aq) + CuSO_4(aq) \rightarrow CuCO_3(s) + Na_2SO_4(aq)$

The state symbols show the salts in solution as (aq) and the precipitate – the insoluble salt – as (s).

In the laboratory

The practical method involves making a precipitate of an insoluble salt by mixing solutions of two soluble salts.

The flow diagram shows how to make an insoluble salt.

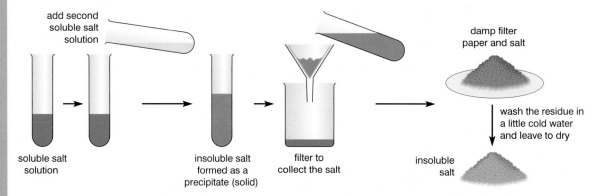

△ Fig. 2.51 Making an insoluble salt.

QUESTIONS

1. What is a precipitation reaction?

2. When preparing an insoluble salt what process is used to separate the insoluble salt from any soluble salts?

3. Why is the insoluble salt washed with a little cold water before it is left to dry?

4. Lead(II) chloride is an insoluble salt and can be prepared by mixing solutions of lead(II) nitrate and sodium chloride.

    a) Write a word equation for the reaction.

    b) Write a balanced symbol equation for the reaction.

Salt is a word commonly used for sodium chloride which we use in cooking. The word salt also refers to a large group of chemical substances which can be formed by the reactions of acids with a variety of alkalis, bases, carbonates and metals. Another salt in common use is magnesium sulfate, often found as $MgSO_4 \cdot 7H_2O$. There are many different uses of this salt. Use what you know about acids, bases and salts to answer the questions below.

1. A student wanted to prepare a sample of magnesium sulfate crystals. The student was provided with dilute sulfuric acid and magnesium carbonate.

   a) Write a word and balanced symbol equation for this reaction.

   b) Produce a risk assessment which the student could use.

   c) List the equipment required for this preparation.

   d) Give the instructions needed for the preparation of the magnesium sulfate.

   e) Unfortunately the student prepared a white powder rather than a crystalline sample. Give reasons which would explain why a powder and not crystals was produced.

   f) Another student suggested he use a titration method to prepare the salt. Why is this not a suitable method to use?

   g) The anhydrous form of magnesium sulfate can be used as a drying agent. Give the formula of the anhydrous form and suggest a reason why it can act as a drying agent.

Developing investigative skills

A student was set the challenge of making a pure, dry sample of hydrated copper(II) sulfate crystals starting from copper(II) oxide and dilute sulfuric acid. (Note: At concentrations between 1.5 mol/dm³ and 0.5 mol/dm³ sulfuric acid is classified as an irritant to skin and eyes. Copper(II) oxide is classified as harmful if swallowed or inhaled, and as an irritant to skin and eyes, and very toxic to aquatic organisms.) The student tackled the challenge in a series of steps:

Step 1

The student put 50 cm³ of dilute sulfuric acid in a beaker and warmed it on a tripod and gauze over a Bunsen burner. The student then added the copper(II) oxide a spatula at a time, stirring to ensure the copper(II) oxide had reacted before adding another spatula.

Step 2

When some copper(II) oxide remained unreacted the student filtered the mixture and collected the copper(II) sulfate solution in an evaporating basin.

Step 3

The student then heated the solution in the evaporating basin until crystallisation point was reached and then allowed the solution to cool and crystallise.

Step 4

The student ran out of time and couldn't finish the challenge.

Demonstrate and describe techniques

❶ In step 1 why did the sulfuric acid need to be heated?

❷ In step 2 what general name is given to the liquid that passes through the filter paper?

❸ In step 3 how do you test for crystallisation point and how do you know when it has been reached?

❹ How could pure and dry crystals be obtained in step 4?

Analyse and interpret data

❺ Write a word equation for the reaction in this experiment.

❻ Write a fully balanced symbol equation for the reaction.

❼ The crystals produced would be hydrated. What does this mean?

❽ What is the chemical formula for hydrated copper(II) sulfate?

Developing investigative skills

A student was using a precipitation method to make a sample of lead(II) sulfate. The student decided to use some sodium sulfate and some lead(II) nitrate, which the student made into solutions by dissolving the solids in water. (Note: sodium sulfate solution is classified as low hazard; lead compounds and solutions are harmful if swallowed or inhaled and very toxic to aquatic organisms.) After mixing the two solution in a beaker the student filtered the mixture and collected the precipitate of lead(II) sulfate on the filter paper.

Demonstrate and describe techniques

❶ Why was the student confident that sodium sulfate would be soluble in water?

❷ What should the student do to ensure that the lead(II) sulfate sample on the filter paper is pure?

Analyse and interpret data

❸ Write a word equation for the reaction used to make the lead(II) sulfate.

❹ Write a fully balanced symbol equation for the reaction.

❺ **a)** In the reaction the student dissolved 10 g of each of solid to make the reacting solutions. Was the lead(II) nitrate or the sodium sulfate in excess? Explain your answer.

b) What mass of lead(II) sulfate would you expect to be formed? Explain your answer and show your working.

c) In the experiment itself 6 g of lead(II) sulfate was produced. Calculate the percentage yield in the reaction. Explain your answer and show your working.

(Relative atomic masses: N = 14, O = 16, S = 32, Na = 23, Pb = 207)

End of topic checklist

An **acid** is a substance that contains replaceable hydrogen atoms which form H^+ ions in water. It is a proton donor.

A **base** is the oxide or hydroxide of a metal.

An **alkali** is a base that is soluble in water and produces OH^- ions. It is a proton acceptor.

A **salt** is formed when the replaceable hydrogen atom(s) of an acid is (are) replaced by a metal.

A **precipitation** reaction is one in which an insoluble salt is formed by mixing two solutions.

The facts and ideas that you should know and understand by studying this topic:

○ Understand acids and bases in terms of proton transfer – an acid is a proton donor and a base is a proton acceptor.

○ Know that metal oxides, metal hydroxides and ammonia can act as bases.

○ Know that the salts made from hydrochloric acid are called chlorides.

○ Know that the salts made from nitric acid are called nitrates.

○ Know that the salts made from sulfuric acid are called sulfates.

○ Be able to predict the products of reactions between dilute hydrochloric, nitric and sulfuric acids and metals, metal oxides and metal carbonates to form salts.

○ Know the general rules for predicting whether an ionic compound is soluble or insoluble in water (see page 169).

○ Be able to describe how a pure, dry sample of a soluble salt can be prepared by adding an insoluble metal, base or carbonate to an acid.

○ Be able to describe experiments to prepare a pure, dry sample of a soluble salt by titration using an acid and an alkali.

○ Be able to describe experiments to prepare an insoluble salt from two soluble reactants, using a precipitation reaction.

○ Be able to prepare a sample of pure, dry hydrated copper(II) sulfate crystals starting from copper(II) oxide.

○ Be able to prepare a sample of pure, dry lead(II) sulfate.

End of topic questions

1. Calcium chloride can be made from calcium oxide and dilute hydrochloric acid.

   a) What type of chemical is calcium oxide? **(1 mark)**

   b) What type of chemical is calcium chloride? **(1 mark)**

   c) Is calcium chloride soluble or insoluble in water? **(1 mark)**

   d) Describe the different stages in the preparation of calcium chloride crystals.
   (4 marks)

   e) Write a fully balanced symbol equation, including state symbols, for the reaction.
   (2 marks)

2. Barium sulfate is an insoluble salt and can be made using a precipitation reaction.

   a) What acid can be used to make barium sulfate? **(1 mark)**

   b) What other chemical could be used to make the barium sulfate? **(1 mark)**

   c) Describe the different stages in the preparation of a dry sample of barium sulfate. **(3 marks)**

   d) Write a fully balanced symbol equation, including state symbols, for the reaction.
   (2 marks)

3. Copy and complete the following equations and include state symbols:

   a) $2KOH(aq) + H_2SO_4(aq) \rightarrow$ _____ + _____ **(2 marks)**

   b) $2HCl(aq) + MgO(s) \rightarrow$ _____ + _____ **(2 marks)**

   c) $2HNO_3(aq) + BaCO_3(s) \rightarrow$ _____ + _____ + _____ **(2 marks)**

   d) $2HCl(aq) + Zn(s) \rightarrow$ _____ + _____ **(2 marks)**

   e) $ZnCl_2(aq) + K_2CO_3(aq) \rightarrow$ _____ + _____ **(2 marks)**

4. A dry sample of potassium chloride can be prepared by titrating hydrochloric acid and potassium hydroxide solution. Describe the stages in this process. **(7 marks)**

Chemical tests

△ Fig. 2.52 These candles are burning with different colours because they contain different metals.

INTRODUCTION

It is important to be able to analyse different substances and identify the different elements or components. The techniques used today are fairly sophisticated, but many of them are based on simple laboratory tests. With improved understanding of the beneficial and harmful properties of chemical substances, it has become more and more important to identify metals and non-metals in chemical processes and in the environment.

KNOWLEDGE CHECK

✓ Understand the nature of the chemical bonding in ionic compounds.
✓ Be familiar with the terms anion and cation.
✓ Know some of the characteristics of Group 1 and Group 7 elements.
✓ Know the order of the common metals that are included in the reactivity series.

LEARNING OBJECTIVES

✓ Be able to describe how to perform a flame test and know the flame colours associated with lithium, sodium, potassium, calcium and copper.
✓ Be able to describe simple tests used to identify hydrogen, oxygen, carbon dioxide, ammonia and chlorine gases.
✓ Be able to describe simple tests for identifying metal and ammonium ions (cations) in solution, and know the outcomes of the tests.
✓ Be able to describe simple tests for identifying non-metal ions (anions) in solution and know the outcomes of the tests.
✓ Describe how to test for the presence of water using anhydrous copper(II) sulfate.
✓ Describe a physical test to show whether a sample of water is pure.

IDENTIFYING METAL IONS (CATIONS)

Ions of metals are **cations** – positive ions – and are found in ionic compounds. There are two ways of identifying metal cations:

- *either* from solids of the compound
- *or* from solutions of the compound.

METAL IONS IN SOLIDS

Flame tests

In a flame test, a piece of nichrome wire is dipped into concentrated hydrochloric acid, then into the solid compound, and then into a blue Bunsen flame. The colour seen in the flame identifies the metal ion in the compound.

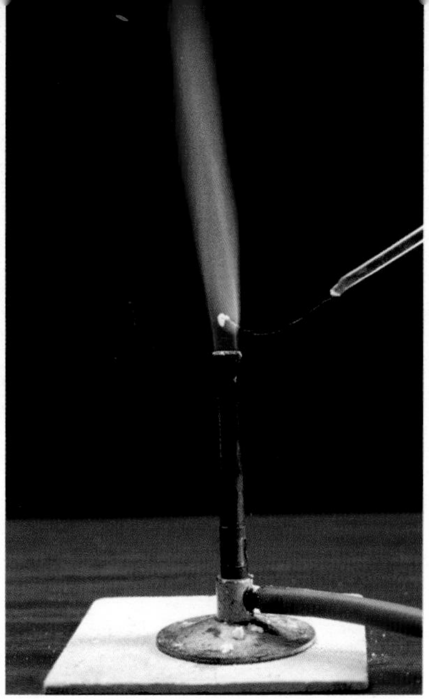

△ Fig. 2.53 The colour of the flame can be used to identify the metal ions present.

Name of ion	Formula of ion	Colour seen in flame
lithium	Li^+	red
sodium	Na^+	yellow
potassium	K^+	lilac (light purple)
calcium	Ca^{2+}	orange-red
copper	Cu^{2+}	blue-green

△ Table 2.10 Colours of ions in a flame.

Tests on solutions in water

Metal ions are found in ionic compounds, so most of them will dissolve in water to form solutions. These solutions can be tested with sodium hydroxide solution to identify the aqueous cation:

Name of ion in solution	Formula	Test	Result
copper(II)	$Cu^{2+}(aq)$	Add sodium hydroxide solution in drops.	Light blue precipitate formed.
iron(II)	$Fe^{2+}(aq)$	Add sodium hydroxide solution in drops.	Green precipitate formed. On standing, changes to reddish brown colour.
iron(III)	$Fe^{3+}(aq)$	Add sodium hydroxide solution in drops.	Reddish brown precipitate formed.

△ Table 2.11 Tests on ions in solution.

REMEMBER

All the reactions with sodium hydroxide solution produce insoluble metal hydroxides: $Cu(OH)_2(s)$, $Fe(OH)_2(s)$, $Fe(OH)_3(s)$

You need to be able to write the formulae for these and their ions.

Remember: copper and iron are **transition metals**, so the Roman numeral tells you the charge on the ion.

The green $Fe(OH)_2$ oxidises in air ($Fe^{2+} \rightarrow Fe^{3+}$) to form the reddish brown $Fe(OH)_3$.

QUESTIONS

1. What colour do calcium ions produce in a flame test?

2. When carrying out a flame test, what sort of wire is used?

3. What test can be used to distinguish between Fe^{2+} and Fe^{3+} ions? What is the result of the test with each ion?

IDENTIFYING AMMONIUM IONS, NH_4^+

The test for the ammonium ion is shown in the diagram.

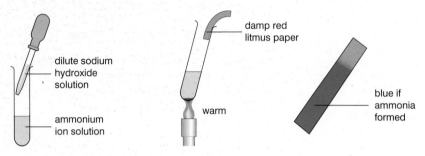

△ Fig. 2.54 Test for the ammonium ion NH_4^+.

The suspected ammonium compound is dissolved in water in a test tube and a few drops of dilute sodium hydroxide are added. The mixture is then warmed over a Bunsen burner and some damp red litmus paper (or universal indicator paper) is placed in the mouth of the test tube. A colour change in the indicator to blue (alkaline) shows that an ammonium compound is present.

IDENTIFYING ANIONS

As with metal cations, negative ions (**anions**) are tested as solids or as solutions.

Testing for anions in solids

The following test for anions in solids applies only to carbonates.

Dilute hydrochloric acid is added to the solid, and any gas produced is passed through lime water. If the lime water goes cloudy/milky, the solid contains a carbonate.

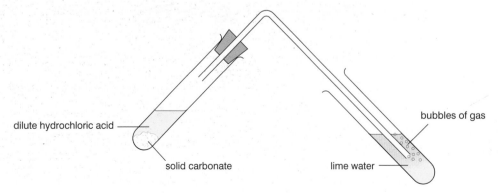

△ Fig. 2.55 Testing for anions in solids.

This reaction is as follows:

	acid	+	carbonate	→	a salt	+	water	+	carbon dioxide
Example	$2\ HCl\ (aq)$	+	$Na_2CO_3(s)$	→	$2\ NaCl(aq)$	+	$H_2O(l)$	+	$CO_2\ (g)$
	hydrochloric acid	+	sodium carbonate		sodium chloride	+	water	+	carbon dioxide
Example	$2\ HCl\ (aq)$	+	$ZnCO_3(s)$	→	$ZnCl_2(aq)$	+	$H_2O(l)$	+	$CO_2(g)$
	hydrochloric acid	+	zinc carbonate		zinc chloride	+	water	+	carbon dioxide

Testing for anions in solution

Many ionic compounds are soluble in water, and so they form solutions that contain anions.

The tests and results used to identify some common anions are shown in the table.

Name of ion	Formula	Test	Result
chloride	$Cl^-(aq)$	To a solution of the halide ions add: 1. dilute nitric acid 2. silver nitrate solution.	white precipitate (of AgCl)
bromide	$Br^-(aq)$		cream precipitate (of AgBr)
iodide	$I^-(aq)$		yellow precipitate (of AgI)
sulfate	$SO_4^{2-}(aq)$	Add: 1. dilute hydrochloric acid 2. barium chloride solution.	white precipitate (of $BaSO_4$)

△ Table 2.12 Tests for anions.

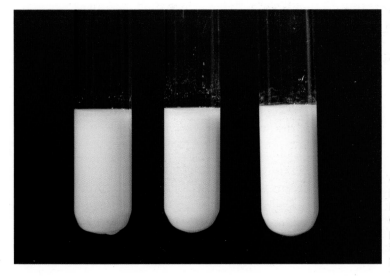

◁ Fig. 2.56 Test for the halide ions.
Left: chloride, white precipitate.
Centre: bromide, pale cream precipitate.
Right: iodide, pale yellow precipitate.

If the silver halides – AgCl, AgBr and AgI, formed as in the table above – are left to stand in daylight for a while, they go dark grey or black. This is because the light reduces them to silver. This darkening in light is the basis of photographic processes that use silver salts on camera film.

QUESTIONS

1. Describe how you would test for an ammonium compound. Give the result of the test.

2. When a carbonate is treated with dilute hydrochloric acid, a gas is given off.

   a) What is the name of the gas?

   b) What is the test for the gas? Give the result of the test.

3. Sodium hydroxide solution is added to solution X and a reddish brown precipitate is formed. What metal ion was present in solution X?

4. A mixture of dilute nitric acid and silver nitrate solution is added to solution Y in a test tube. A white precipitate forms. What anion is present in solution Y?

IDENTIFYING GASES

Many chemical reactions produce a gas as one of the products. Identifying the gas is often a step towards identifying the compound that produced it in the reaction (see Testing for anions on pages 180 and 181).

Gas	Formula	Test	Result of test
hydrogen	H_2	Put in a lighted splint (a flame).	'Pop' or 'squeaky pop' heard (flame usually goes out).
oxygen	O_2	Put in a glowing splint.	Splint relights, producing a flame.
carbon dioxide	CO_2	Pass gas through lime water.	Lime water goes cloudy/milky.
chlorine	Cl_2	Put in a piece of damp blue litmus paper.	Paper goes red then white (bleached).
ammonia	NH_3	Put in a piece of damp red litmus or universal indicator paper.	Paper goes blue.

△ Table 2.13 Tests for gases.

Remember the following:

Carbon dioxide: cloudiness with lime water is insoluble calcium carbonate. If carbon dioxide continues to be passed through, the cloudiness disappears: $CaCO_3(s)$ is changed to soluble calcium hydrogen carbonate, $Ca(HCO_3)_2(aq)$.

Chlorine: the gas is acidic, but also a bleaching agent.

Ammonia: the only alkaline gas.

QUESTIONS

1. What is the name of a gas that is alkaline?

2. What is the name of a gas that supports combustion?

3. What is the name of a gas that acts as a bleach?

SCIENCE IN CONTEXT

FLAME EMISSION SPECTROSCOPY

This modern analytical technique is like a sophisticated flame test. A sample of a material is squirted into a flame as a gas or sprayed solution. The heat from the flame evaporates the solvent and breaks chemical bonds to create free atoms. The thermal energy also excites the atoms into electronic states that emit light when they return to the ground electronic state. Each element emits light at a characteristic wavelength, which is dispersed by a grating or prism and detected in the spectrometer.

This technique is frequently applied to regulate alkali metals used in the pharmaceutical industry.

Qualitative analysis can be a useful tool for analytical chemists and forensic scientists. It enables them to identify both cations and anions in many chemical substances. Use your knowledge of the tests for ions and gases to help you answer questions below.

1. Metallic ions can be identified using flame tests or sodium hydroxide solution.

   a) In a flame test it is important to ensure that the wire used is not contaminated with sodium ions. How can you do this?

   b) Sodium hydroxide is useful as it forms coloured precipitates with many metallic ions.

   Copy and complete the table with the names of three cations which give coloured precipitates.

Name of cation	Colour of precipitate

   c) Halide ions can be identified using nitric acid and silver nitrate. Give a general ionic equation to show the formation of the precipitate of the halide ion.

   d) When testing for sulfate ions why is it important to add dilute hydrochloric acid before adding barium chloride?

2. A forensic scientist has been provided with a small sample of a blue compound which is suspected to be copper(II) sulfate and a white compound that is suspected to be sodium carbonate. Devise a series of tests that could be followed to identify the ions. Indicate in your plan the expected results if the samples are to be positively identified.

A SIMPLE CHEMICAL TEST FOR WATER

Water turns **anhydrous** copper(II) sulfate from white to blue. When anhydrous copper(II) sulfate meets water, water of crystallisation is added, and this makes blue crystals.

$$CuSO_4(s) \quad + \quad 5H_2O(l) \quad \rightarrow \quad CuSO_4.5H_2O(s)$$

This test will always show the presence of water, but it will not show whether the water is pure.

△ Fig. 2.57 The chemical test for water – adding water to anhydrous copper(II) sulfate.

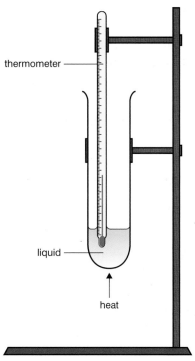

thermometer

liquid

heat

△ Fig. 2.58 Apparatus for testing if water is pure.

A PHYSICAL TEST TO SHOW IF WATER IS PURE

To test if water is pure, test its boiling point with an accurate thermometer. For example, pure water has a boiling point of $100\,°C$ at normal pressure but a solution of calcium chloride in water has a boiling point of $102\,°C$.

QUESTIONS

1. What does the word 'anhydrous' mean?

2. Anhydrous copper(II) sulfate can be used to test for the presence of water. What test can be used to show if a liquid is pure water?

End of topic checklist

An **anion** is a negatively charged ion (during electrolysis it moves to the anode).

A **cation** is a positively charged ion (during electrolysis it moves to the cathode).

An **anhydrous** salt is one that does not contain **water**.

A **hydrated** salt is one that does contain water of crystallisation.

The facts and ideas that you should know and understand by studying this topic:

○ Be able to describe a flame test using a nichrome wire, concentrated hydrochloric acid and a sample of the solid.

○ Know the flame colours produced by the following metal ions: Li^+; Na^+; K^+; Ca^{2+} and Cu^{2+}.

○ Be able to describe the test for the ammonium ion, NH_4^+, using sodium hydroxide solution and know what indicates a positive result.

○ Be able to describe the test for Cu^{2+}, Fe^{2+} and Fe^{3+} ions using sodium hydroxide solution and know the precipitate colour for each ion.

○ Be able to describe the test for Cl^-, Br^- and I^- ions using dilute nitric acid and silver nitrate and know the colours of the different precipitates.

○ Be able to describe the test for SO_4^{2-} ions using dilute hydrochloric acid and barium chloride solution and know what colour indicates a positive result.

○ Be able to describe the test for CO_3^{2-} ions using dilute hydrochloric acid and know what indicates a positive result.

○ Be able to describe the test for the gases hydrogen, oxygen, carbon dioxide, ammonia and chlorine and know what indicates a positive result in each case.

○ Know that anhydrous copper(II) sulfate can be used to identify the presence of water.

○ Be able to describe a physical test to show whether a sample of water is pure.

End of topic questions

1. How would you use a flame test to distinguish between a compound containing sodium and a compound containing potassium? You should:

   a) Describe how a flame test is performed. **(2 marks)**

   b) State the different results for the two compounds. **(2 marks)**

2. What ions are likely to be present in the following compounds?

   a) Solution X forms a pale blue precipitate when sodium hydroxide solution is added. **(1 mark)**

   b) Solution Y forms no precipitate when sodium hydroxide solution is added but produces a strong-smelling, alkaline gas when the mixture is heated. **(1 mark)**

   c) Solution Z forms an orange brown precipitate when sodium hydroxide solution is added. **(1 mark)**

3. Copy and complete the table below which is about the identification of gases. **(3 marks)**

Gas	Test	Observations
Chlorine	Damp universal indicator paper	
	Bubble through lime water	White precipitate or suspension forms
Hydrogen		Burns with a 'pop'

4. A white powder is labelled as 'Lithium carbonate'. What two tests could be performed to show that it contains lithium carbonate? **(4 marks)**

5. How would you test a solid to identify the presence of each of the ions shown below?

   a) the sulfate ion, SO_4^{2-} **(3 marks)**

   b) the iodide ion, I^- **(3 marks)**

6. Describe how you could perform a test on a liquid to detect the presence of water. What would you observe if water was present? **(2 marks)**

7. Describe a test that can be used to show whether a sample of water is pure. **(2 marks)**

Exam-style questions
Sample student answers

Question 1

Lithium (Li), sodium (Na) and potassium (K) are in Group 1 of the Periodic Table.

a) These elements have similar chemical properties.

Explain why, using ideas about electronic structures.

All the elements have one electron in their outer shell. ✔ (1)

b) Lithium reacts with water to form a solution of lithium hydroxide and a colourless gas. During this reaction the temperature of the water increases.

i) What is the name of the colourless gas produced?

hydrogen ✔ (1)

ii) Why does the temperature of the water increase?

The reaction between lithium and water is rapid. ✘ (1)

iii) Describe what you would observe if a small piece of sodium is added to water.

The sodium forms sodium hydroxide and hydrogen gas is given off. ✘ *The reaction would be more rapid than the lithium reaction.* ✔ (2)

iv) Caesium (Cs) is another Group 1 metal.

Is caesium more or less reactive than lithium? Give a reason for your answer.

More reactive, because reactivity in Group 1 increases down the group. ✔ (1)

EXAMINER'S COMMENTS

a) The mark would have been given for saying that all the elements have the same number of electrons in their outer shell. The candidate has gone further and correctly stated that they all have one electron in the outer shell.

b) (i) The correct answer has been given.

(ii) This answer lacks precision. The mark would be awarded for either saying that the reaction produced heat or that the reaction was exothermic.

(iii) The candidate has not scored both marks. Apart from stating that the reaction would be more rapid than that with lithium, observations have not been given. The products have been correctly named but these are not what you would observe. Marks would be awarded for: fizzing/ effervescence/ bubbles (of gas), the sodium floats/ moves around on the water, forms a ball/disappears.

(iv) The correct answer and explanation have been given.

c) The correct flame colour has been given.

Exam-style questions cont.

c) The presence of a sodium compound can be tested using a flame test.

What is the flame test colour for a sodium compound?

yellow ✓

(1)

(Total 7 marks)

Question 2

Solutions of lead(II) nitrate and potassium iodide react together to make the insoluble substance lead(II) iodide.

The equation for the reaction is

$$Pb(NO_3)_2(aq) + 2KI(aq) \rightarrow 2KNO_3(aq) + PbI_2(s)$$

An investigation was carried out to find how much precipitate formed with different volumes of lead(II) nitrate solution.

A student measured out 15 cm³ of potassium iodide solution using a measuring cylinder.

He placed this solution in a clean boiling tube.

Using a clean measuring cylinder, he measured out 2 cm³ of lead(II) nitrate solution (of the same concentration, in mol/dm³, as the potassium iodide solution). He added this to the potassium iodide solution.

A cloudy yellow mixture formed and this was left to settle.

The student then measured the height (in cm) of the precipitate using a ruler.

The student repeated the experiment using different volumes of lead(II) nitrate solution. The graph shows the results obtained.

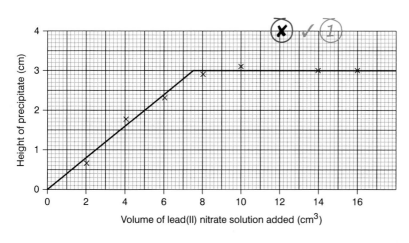

Exam-style questions cont.

a) i) On the graph, circle the point which seems to be anomalous. **(1)**

ii) Explain two things that the student may have done in the experiment to give this anomalous result.

Precipitate not settled ✓ ①
Because not left long enough ✓ ① **(4)**

iii) Why must the graph line go through (0,0)?

Cannot have a precipitate
if no lead nitrate added yet. ✓ ① **(1)**

b) Suggest a reason why the height of the precipitate stops increasing.

No more potassium
iodide left to react. ✓ ① **(1)**

c) i) How much precipitate has been made in the tube?

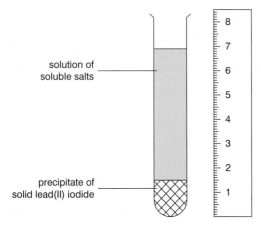

solution of soluble salts

precipitate of solid lead(II) iodide

1.5 cm ✓ ① **(1)**

ii) Use the graph to find the volume of lead(II) nitrate solution needed to make this amount of precipitate.

2.9 cm ✗ **(1)**

d) After he had plotted the graph, the student decided he should obtain some more results.

i) Suggest what volumes of lead(II) nitrate solution he should use.

Between 6 cm³ and 10 cm³ ✓ ① **(1)**

Exam-style questions cont.

ii) Explain why he should use these volumes.

Need to know exactly where the graph levels off ✓ ①
(1)

e) Suggest a different method for measuring the amount of precipitate formed. This method must not be based on the height of the precipitate.

Filter ✓ ① *off each precipitate and weigh it* ✓ ①
(4)

(Total 15 marks)

Question 3

Use the Periodic Table to help you answer this question.

a) State the symbol of the element with atomic number 14. ....
(1)

b) State the symbol of the element that has a relative atomic mass of 32. ....
(1)

c) State the number of the group that contains the alkali metals. ....
(1)

d) Which group contains elements whose atoms form ions with a 2$^+$ charge? ....
(1)

e) Which group contains elements whose atoms form ions with a 1$^-$ charge? ....
(1)

(Total 5 marks)

Question 4

Three of the elements in Group 7 of the Periodic Table are chlorine, bromine and iodine.

a) Chlorine has an atomic number of 17. What is the electron configuration of chlorine? ....
(1)

b) How many electrons will be in the outer shell of the bromine atom? ....
(1)

c) Bromine reacts with hydrogen to form hydrogen bromide. The chemical equation for the reaction is:

$Br_2(g) + H_2(g) \rightarrow 2HBr(g)$

What is the colour change during the reaction? ....
(1)

d) Hydrogen iodide and hydrogen chloride have similar properties.

   i) A sample of hydrogen iodide is dissolved in water. A piece of universal indicator paper is dipped into the solution. State, with a reason, the final colour of the universal indicator paper.   **(2)**

   ii) A sample of hydrogen iodide is dissolved in methyl benzene. A piece of universal indicator paper is dipped into the solution. State, with a reason, the final colour of the universal indicator paper.   **(2)**

(Total 7 marks)

Question 5

The reactivity of metals can be compared by comparing their reactions with dilute sulfuric acid. Pieces of zinc, iron and magnesium of identical size are added to separate test tubes containing this acid.

a) What order of reactivity would you expect? Put the most reactive metal first.   **(1)**

b) Write a word equation for the reaction between magnesium and dilute sulfuric acid.   **(1)**

c) Write a fully balanced equation for the reaction in **b)**.   **(1)**

d) Name a metal that does not react with dilute sulfuric acid.   **(1)**

e) What other reaction could be used to compare the reactivity of metals?   **(1)**

(Total 5 marks)

Question 6

Read the following instructions for the preparation of hydrated nickel(II) sulfate ($NiSO_4.7H_2O$), then answer the questions that follow.

1. Put 25 cm^3 of dilute sulfuric acid in a beaker.

2. When the nickel carbonate has dissolved, add a little more nickel carbonate. Continue in this way until nickel carbonate is in excess.

3. Filter the mixture into a clean beaker.

Exam-style questions cont.

4. Make the hydrated nickel(II) sulfate crystals from the nickel(II) sulfate solution.

The equation for the reaction is

$$NiCO_3(s) + H_2SO_4(aq) \rightarrow NiSO_4(aq) + CO_2(g) + H_2O(l)$$

a) What piece of apparatus would you use to measure out 25 cm³ of sulfuric acid? **(1)**

b) Why is the nickel(II) carbonate added in excess? **(1)**

c) When nickel(II) carbonate is added to sulfuric acid, there is fizzing. Explain why. **(1)**

d) Draw a diagram to describe step 4. You must label your diagram. **(3)**

e) After filtration, which one of the following describes the nickel(II) sulfate in the beaker?

Select the correct answer.

crystals   filtrate   precipitate   water   **(1)**

f) Explain how you would obtain pure dry crystals of hydrated nickel(II) sulfate from the solution of nickel(II) sulfate. **(2)**

g) When hydrated nickel(II) sulfate is heated gently in a test tube, it changes colour from green to white.

   i) Complete the symbol equation for this reaction.

$$NiSO_4.7H_2O(s) \rightleftharpoons NiSO_4(s) +$$ **(1)**

   ii) What does the sign $\rightleftharpoons$ mean? **(1)**

h) How can you obtain a sample of green nickel(II) sulfate starting with white nickel(II) sulfate? **(1)**

(Total 12 marks)

Question 7

The ions present in ionic compounds can be identified using simple tests. The table below shows the flame colours of three cations.

Cation	Flame test colour
lithium	red
sodium	yellow
calcium	orange-red

The following table shows the results of three tests that can be used to identify some anions in solution:

Anion	Result of adding dilute nitric acid	Result of adding universal indicator solution
carbonate	effervescence	blue
hydrogen carbonate	effervescence	dark green
hydrogen sulfate	no effervescence	red
hydroxide	no effervescence	blue
sulfate	no effervescence	green

Two ionic compounds, **X** and **Y**, containing a combination of the ions shown above were analysed.

a) Describe how you would carry out a flame test on a solid compound. .... (2)

b) Compound **X** gave a yellow flame test and produced effervescence when dilute nitric acid was added. Suggest two possible names for **X**. .... (2)

c) Compound **Y** gave a red flame test and universal indicator turned blue. Give the names of the possible compounds **Y** could be. .... (2)

(Total 6 marks)

Question 8

This question is about the metals magnesium and lead.

a) Magnesium reacts with lead(II) oxide forming magnesium oxide and lead.

    i) What is the name given to this type of reaction? .... (1)

    ii) Write a balanced symbol equation, including state symbols, for this reaction. ....(2)

    iii) Which metal is more reactive, magnesium or lead? Give a reason for your answer. .... (1)

b) When the magnesium reacts with the lead(II) oxide:

    i) Name the substance that has been oxidised. .... (1)

    ii) Name the substance that has been reduced. .... (1)

    iii) Name the substance that is acting as an oxidising agent. .... (1)

    iv) Name the substance that is acting as a reducing agent. .... (1)

c) Magnesium will also react with lead(II) nitrate solution, $Pb(NO_3)_2(aq)$, forming lead.

    i) Write a balanced symbol equation, including state symbols, for this reaction. .... (2)

    ii) Name a metal that will also react with lead(II) nitrate solution to form lead. .... (1)

    iii) Name a metal that will not react with lead(II) nitrate solution to form lead. .... (1)

(Total 12 marks)

Modern physical chemistry has its origins in the chemistry of the 19th century. This category is not as clearly defined a category of chemistry, but the term is still a useful description of this branch of science. Physical chemistry focuses on chemical processes from the 'macro-level' – where properties can be observed – rather than at the 'micro-level' of individual atoms, molecules and ions. However, observed physical properties can still be explained in terms of what the atoms, molecules or ions do.

In this section you will explore the potential for chemical reactions to generate significant amounts of heat energy, as well as some strange reactions that seem to absorb energy and make everything cooler. The applications of the first type of these reactions are readily recognisable in the chemistry of fuels. The speed or rate of chemical reactions will also be explored, together with chemists' strategies to try and control them. Finally, you will learn about reactions that go from reactants to products and then back again. These are a particular challenge when chemists want to make a product that does not turn back into the reactants that made it!

STARTING POINTS

1. How many non-renewable fuels can you name? What products do they form when they burn?

2. Give an example of a very rapid, almost instantaneous, chemical reaction. Give an example of a very slow one.

3. Have you heard of the process of neutralisation? If so, in what context?

4. What is a catalyst? Do you know of any examples where catalysts are used in everyday life?

5. Can you give any examples of processes that are reversible?

SECTION CONTENTS

a) Energetics

b) Rates of reaction

c) Reversible reactions and equilibria

d) Exam-style questions

3
Physical chemistry

△ Physical chemistry deals with properties that can be observed.

Energetics

INTRODUCTION

When chemicals react together, the reactions cause energy changes. This is obvious when a fuel is burnt and heat energy is released into the surroundings. Heat changes in other reactions may be less dramatic but they still take place. A knowledge of chemical bonding can really help to understand how these energy changes occur.

△ Fig. 3.1 Fireworks, a carefully controlled chemical reaction.

KNOWLEDGE CHECK

✓ Know that atoms in molecules are held together by covalent bonds.
✓ Know that many common fuels are organic compounds called alkanes.
✓ Be able to write and interpret balanced chemical equations.

LEARNING OBJECTIVES

✓ Know that chemical reactions in which heat energy is given out are described as exothermic, and those in which heat energy is taken in are described as endothermic.
✓ Be able to describe simple calorimetry experiments for changes such as combustion, displacement, dissolving and neutralisation.
✓ Be able to calculate the heat energy change from a measured temperature change using the expression $Q = mc\Delta T$.
✓ Be able to calculate the molar enthalpy change (ΔH) from the heat energy change, Q.
✓ Be able to draw and explain energy level diagrams to represent exothermic and endothermic reactions.
✓ Know that bond-breaking is an endothermic process and that bond-making is an exothermic process.
✓ Be able to use bond energies to calculate the enthalpy change during a chemical reaction.
✓ Be able to investigate temperature changes accompanying some of the following types of change:
  • salts dissolving in water
  • neutralisation reactions
  • displacement reactions

ENERGY CHANGES IN CHEMICAL REACTIONS

In most reactions, energy is transferred to the surroundings and the temperature goes up. These reactions are **exothermic**. In a minority of cases, energy is absorbed from the surroundings as a reaction takes place and the temperature goes down. These reactions are **endothermic**.

For example, when magnesium ribbon is added to dilute hydrochloric acid, the temperature of the acid increases – the reaction is exothermic. In contrast, when sodium hydrogencarbonate is added to hydrochloric acid, the temperature of the acid decreases – the reaction is endothermic.

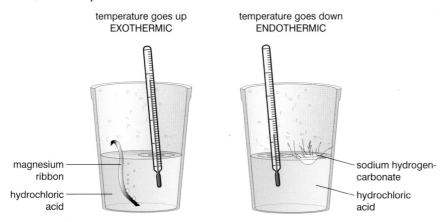

△ Fig. 3.2 Measuring energy changes in a reaction.

Energy changes in reactions like these can be measured in a polystyrene cup as a calorimeter. If a lid is added to the cup, very little energy is transferred to the air and quite accurate results can be obtained. A simple equation is used to calculate the energy change.

energy transferred to the solution	=	**mass of the solution**	×	**specific heat capacity of water**	×	**change in temperature**
kilojoules, kJ (or J)		kg (or g)		4.2k J/kg/°C (or 4.2 J/g/°C)		°C

In short, energy change = $m \times c \times \Delta T$

where c is the specific heat capacity of the substance being heated, in this case water.

Note: In this calculation it is assumed that all liquids or solutions have the same specific heat capacity as water and the same density (so 1000 cm³ has a mass of 1 kg, 1 cm³ has a mass of 1 g).

WORKED EXAMPLE

1 g of magnesium ribbon was added to 200 g of 1 mol/dm³ hydrochloric acid in a polystyrene beaker. When the magnesium had completely reacted and none was left, the temperature of the acid had risen by 30 °C. Calculate the energy change for the reaction.

Equation: energy change = mass of hydrochloric acid × 4.2 × temperature change

Substitute values: energy change = 200 × 4.2 × 30

Calculate: energy change = 25 200 J or 25.2 kJ for 1 g of magnesium

1. What is an exothermic reaction?

2. What is an endothermic reaction?

3. Why do polystyrene cups make good calorimeters for measuring energy changes in some chemical reactions?

4. Adding 2.5 g of sodium hydroxide to 100 g of water in a polystyrene cup produced a temperature rise of 6 °C. Calculate the energy change.

All reactions involving the combustion of fuels are exothermic. The energy transferred when a fuel burns can be measured using **calorimetry**, as shown in the diagram.

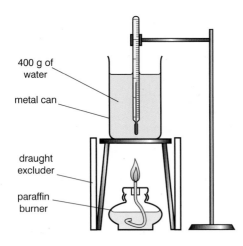

400 g of water

metal can

draught excluder

paraffin burner

◁ Fig. 3.3 Measuring the energy produced on burning a liquid fuel.

The rise in temperature of the water is a measure of the energy transferred to the water. This technique will not give a very accurate answer because much of the energy will be transferred to the surrounding air. Nevertheless, it can be used to compare the energy released by the same amounts of different fuels.

WORKED EXAMPLE

2.0 g of paraffin were burned in a spirit burner under a metal can containing 400 cm^3 of water. The temperature of the water rose from 20 °C to 70 °C. Calculate the energy produced by the paraffin in J/g and kJ/g.

Equation:	heat change = mass of water × 4.2 × temperature change
Substitute values:	heat change = 400 × 4.2 × 50
Calculate:	heat change = 84 000 J per 2 g of paraffin
	= 42 000 J/g
	= 42 kJ/g

Note: This calculation assumes that all the heat energy is transferred to the water and none is retained by the metal can.

ENTHALPY CHANGE

The heat energy in chemical reactions is called enthalpy. The **enthalpy change** is given the symbol ΔH. The enthalpy change for a particular reaction is shown at the end of the balanced equation. The units are (k)J/mol.

Units are important. If in the calculation you use g for mass of water, then the change is in J, but using kg for the mass gives the enthalpy change in kJ.

The enthalpy change shown at the end of a balanced equation will be the **molar enthalpy change**. This is the energy change if the molar quantities shown in the equation react together. In many reactions the exact molar quantities are not used, but the molar enthalpy change can be calculated from the actual quantities used.

WORKED EXAMPLE

A spirit burner was half filled with liquid heptane (C_7H_{16}), weighed and put under a beaker containing 200 cm^3 of water on a tripod. The temperature of the water was recorded and then the burner was lit. The water in the beaker was carefully stirred with a thermometer. When the temperature had risen exactly 20 °C the flame was extinguished and the burner was re-weighed. It had lost 0.5 g in mass. Use these results to calculate the molar enthalpy change for the reaction. (A_r: H = 1; C = 12; 1 mole of heptane = 100 g)

$$C_7H_{16}(l) + 11O_2(g) \rightarrow 7CO_2(g) + 8H_2O(l) \qquad \Delta H = ?$$

Equation: enthalpy change = mass of water × SHC × rise in temperature

Substitute values: enthalpy change = 200 × 4.2 × 20
Calculate: enthalpy change = 16 800 J for 0.5 g heptane

molar enthalpy change = 16 800 × 100/0.5 = 3 360 000 J

molar enthalpy change = 3360 kJ/mol

QUESTIONS

1. Measuring energy changes when burning fuels in a liquid burner does not give very accurate results. Why do you think this is?

2. Burning 0.10 g of corn puffs raises the temperature of 30 cm^3 of water by 15 °C. Calculate how much energy could be obtained from 1 g of corn puffs.

3. Burning 0.5 g of ethanol (C_2H_5OH) raises the temperature of 200 cm^3 of water by 14 °C. Calculate the energy released when 1 mole of ethanol burns. (A_r: H = 1; C = 12; O = 16)

ENERGY PROFILES AND ΔH

Energy level diagrams show the enthalpy difference between the reactants and products.

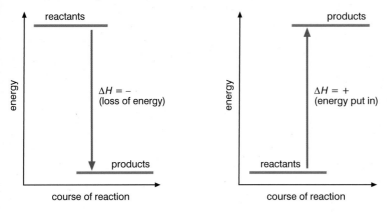

△ Fig. 3.4 Energy level diagrams for exothermic and endothermic reactions.

An exothermic reaction. Energy is being lost to the surroundings. ΔH is negative.

An endothermic reaction. Energy is being absorbed from the surroundings. ΔH is positive.

All ΔH values should have a + or − sign in front of them to show if they are exothermic or endothermic.

Activation energy is the minimum amount of energy required for a reaction to occur. This diagram shows the activation energy of a reaction.

The energy profile can now be completed as shown. The reaction for this profile is exothermic, with ΔH negative.

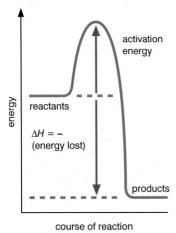

◁ Fig. 3.5 An energy profile for an exothermic reaction.

QUESTIONS

1. The molar enthalpy change for a reaction is positive. Is the reaction endothermic or exothermic?

2. On an energy profile, what is the name given to the minimum amount of energy required for a reaction to occur?

Developing investigative skills

A student wanted to calculate the molar enthalpy change for the neutralisation reaction between sodium hydroxide and hydrochloric acid.

$NaOH(aq) + HCl(aq) \rightarrow NaCl(aq) + H_2O(l)$

The student decided to use a simple calorimetric method.

The student put 50 cm³ of 1 mol/dm³ sodium hydroxide (NaOH) solution in a large polystyrene cup. The student took the temperature of the solution. They then measured the temperature of 50 cm³ of 1 mol/dm³ hydrochloric acid (HCl) in a conical flask. Carefully but quickly the student added the hydrochloric acid solution and stirred the resulting solution with the thermometer. The student recorded the highest temperature reached. The results are shown in Table 3.1.

Volume of 1 mol/dm³ sodium hydroxide solution	50 cm³
Volume of 1 mol/dm³ hydrochloric acid solution	50 cm³
Initial temperature of sodium hydroxide	18 °C
Initial temperature of hydrochloric acid	18 °C
Final temperature of the mixture	25 °C

△ Table 3.1 Results of experiment.

Devise and plan

❶ What apparatus do you think was used to measure out the volumes of sodium hydroxide and hydrochloric acid?

Demonstrate and describe techniques

❷ 1 mol/dm³ sodium hydroxide solution is corrosive. What safety precautions should have been used when carrying out the experiment?

❸ How could the loss of heat energy from the polystyrene cup have been reduced during the experiment?

Analyse and interpret

❹ Calculate the enthalpy change that occurred in the 100 cm³ of the mixture.

❺ Use your answer in 4 to calculate the molar enthalpy change (ΔH) for the reaction. (Do not forget to give your answer a positive or negative sign.)

Evaluate data and methods

❻ How would you have changed the calculation if the temperatures of the sodium hydroxide and hydrochloric acid before mixing were different?

❼ List some possible sources of error in the experiment. Which do you think would be the greatest?

WHERE DOES THE ENERGY COME FROM?

The reaction that occurs when burning fuel can be considered to take place in two stages. In the first stage the covalent bonds between the atoms in the fuel molecules and the oxygen molecules are broken. In the second stage the atoms combine and new covalent bonds are formed. For example, in the combustion of propane:

propane	+	oxygen	→	carbon dioxide	+	water
$C_3H_8(g)$	+	$5O_2(g)$	→	$3CO_2(g)$	+	$4H_2O(l)$

$\Delta H = -2202$ kJ/ mol

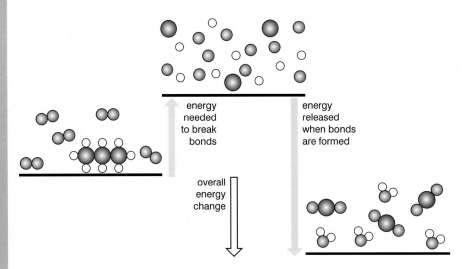

△ Fig. 3.6 Energy changes in an exothermic reaction.

Stage 1: Energy is needed (absorbed from the surroundings) to break the bonds. This process is endothermic.

Stage 2: Energy is released (transferred to the surroundings) as the bonds form. This process is exothermic.

The overall reaction is exothermic because forming the bonds releases more energy than is needed initially to break the bonds. A simplified energy level diagram showing the exothermic nature of the reaction is shown.

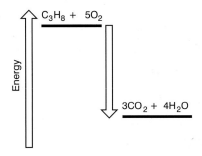

△ Fig. 3.7 Energy changes involved in burning propane.

Propane belongs to a homologous series (family of compounds) called the alkanes. The formulae of the first members of the series are shown in Table 3.2. As shown in the table, the molar enthalpies of combustion increase significantly from methane to hexane, as the number of C–C and C–H bonds increases. So, although as the series is descended the number of C–C and C–H bonds that must be broken increases (absorbing progressively more energy), this is more than counteracted by the progressively more energy released on forming more C=O and O–H bonds.

Alkane	Molar enthalpy of combustion (kJ/mol)
methane CH_4	– 882
ethane C_2H_6	– 1542
propane C_3H_8	– 2202
butane C_4H_{10}	– 2877
pentane C_5H_{12}	– 3487
hexane C_6H_{14}	– 4141

Δ Table 3.2 Molar enthalpy of combustion of alkanes.

REMEMBER

In an exothermic reaction, the energy released on forming new bonds is greater than that needed to break the old bonds.

In an endothermic reaction, more energy is needed to break bonds than is released when new bonds are formed. The energy changes in endothermic reactions are usually relatively small.

The molar enthalpy change figures given in the table for the alkanes are calculated when 1 mole of each alkane is completely burned in a plentiful supply of oxygen to form carbon dioxide and water.
The increase in molar enthalpy change from one alkane to the next is almost constant due to the extra CH_2 unit in the molecule.

BOND ENERGY CALCULATIONS

Every covalent bond has a particular amount of energy needed to break it. This is the same as the amount of energy given out when it is made. This is the **bond energy** and its units are kJ/mol.

Table 3.3 shows some values of average bond energies.

Bond	C–C	C–H	O=O	H–H	H–O	C=O	Cl–Cl	H–Cl
Average bond energy (kJ/mol)	348	413	498	436	464	745	242	431

Δ Table 3.3 Some average bond energies.

WORKED EXAMPLE

Calculate the enthalpy change for the reaction between hydrogen and chlorine:

$$H_2 + Cl_2 \rightarrow 2HCl \qquad\qquad \text{i.e. } H\text{–}H + Cl\text{–}Cl \rightarrow 2 \times H\text{–}Cl$$

What does the sign of the energy change tell you about the reaction?

H–H	+	Cl–Cl	→	2	×	H–Cl
436 kJ /mol		242 kJ/mol		2	×	431 kJ/mol

total for bonds = +678 kJ/mol   total for bonds = –862 kJ/mol
(endothermic because bond   (exothermic because bond making)
breaking)

Overall difference –862 kJ/mol
                    +678 kJ/mol
                    —————————
                    –184 kJ/mol

Answer: $\Delta H = -184$ kJ/mol

It is an exothermic reaction (ΔH negative).

Summary of method:

1. Total all the bonds on the left and allocate a + sign.
2. Total all the bonds on the right and allocate a – sign.
3. Find the difference between the two values. Remember the sign (+ or –).
4. State if exothermic (–) or endothermic (+).

REMEMBER

In bond energy calculation questions you must identify the sign of the answer and link it to exothermic (–) or endothermic (+). You will gain extra credit for linking 'exothermic' to more energy being released by bond making than used in bond breaking, and the opposite for 'endothermic'.

QUESTIONS

1. What does the sign of ΔH indicate about a reaction?

2. Is energy needed or released when bonds are broken?

3. In an endothermic reaction is more or less energy needed to break the bonds than is recovered when bonds are formed?

4. What units are used for bond energy values?

5. Calculate the energy change when 2 moles of hydrogen molecules react with 1 mole of oxygen molecules to make 2 moles of water. Use the bond energy values given on page 205).

HOW COMMON ARE ENDOTHERMIC REACTIONS?

Almost all chemical reactions in which simple compounds or atoms react to make up new compounds are exothermic. One exception is the formation of nitrogen oxide (NO) from nitrogen and oxygen. Overall energy is needed to create this compound, with less energy being released on forming bonds than was needed to break the bonds initially. Nitrogen oxide is often created in lightning storms. The lightning provides enough energy to split the nitrogen and oxygen molecules before the atoms combine to form nitrogen oxide.

△ Fig. 3.8 These plants are making food by photosynthesis, an endothermic reaction.

$$N_2(g) + O_2(g) \rightarrow 2NO(g) \qquad \Delta H \text{ positive}$$

Another exception is photosynthesis. Plants use energy from sunlight to convert carbon dioxide and water into glucose and oxygen.

$$6CO_2(g) + 6H_2O(l) \rightarrow C_6H_{12}O_6(aq) + 6O_2(g) \qquad \Delta H \text{ positive}$$

'Cold packs', which you can buy in some countries, can be used to help you keep cool. Usually you have to bend them to break a partition inside and allow two substances to mix. They will then stay cold for an hour or more. However, it may not be an endothermic reaction that is working in the cold pack. Dissolving chemicals like urea or ammonium nitrate in water causes the temperature of the water to fall, but dissolving is a physical change, not a chemical change. Whether it is an endothermic reaction or not is the manufacturer's secret!

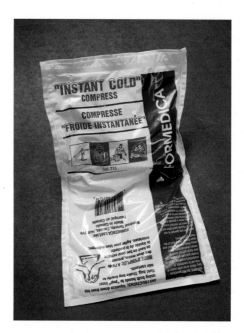

△ Fig. 3.9 A cold pack.

Fuels are substances that react with oxygen to release useful energy. Hydrogen is often considered to be an environmentally friendly alternative to fossil fuels. It is useful to be able to calculate how much heat energy can be given out by a chemical reaction, by using bond energies.

1. When hydrogen gas burns in oxygen the only product is water.

   a) Write a balanced symbol equation for this reaction.

   b) The reaction is exothermic. Explain what this means.

   c) Draw and label an energy profile to show that this reaction is exothermic.

   d) Using the bond energies given in the table below, show that the overall energy change is 243 kJ/mol of hydrogen burned.

Bonds	Average bond energy (kJ/mol)
O=O	498
H–H	436
O–H	464

2. Hydrogen is a non-polluting clean fuel as water is the only product.

   Suggest reasons why this is a fuel which is not yet widely used.

End of topic checklist

An **exothermic reaction** is one in which energy is transferred to the surroundings.

An **endothermic reaction** is one in which energy is taken in from the surroundings.

The **molar enthalpy change** (ΔH) is the heat energy change when the molar quantities shown in the chemical equation react together.

The facts and ideas that you should know and understand by studying this topic:

○ Know that chemical reactions in which heat energy is given out are described as exothermic.

○ Know that chemical reactions in which heat energy is taken in are described as endothermic.

○ Be able to describe how a polystyrene cup can be used to carry out simple calorimetric experiments to determine energy changes (such as in dissolving solids or displacement or neutralisation reactions).

○ Be able to describe how a copper calorimeter can be used to measure the energy change when a fuel is burned (combustion).

○ Understand how to calculate the heat energy change in a calorimetry experiment using the equation:

heat energy change = mass of water (g) × 4.2 (J/g/°C) × change in temperature (°C).

○ Understand that the heat energy change in a reaction is also called the enthalpy change.

○ Understand that ΔH is used to represent the molar enthalpy change for a reaction.

○ Know that ΔH for exothermic reactions is negative and that ΔH for endothermic reactions is positive.

○ Understand how to calculate molar enthalpy change from heat energy change.

○ Understand how energy level diagrams can be used to show the difference between exothermic and endothermic reactions.

○ Understand that the breaking of bonds is endothermic and that the making of bonds is exothermic.

○ Be able to use average bond energies to calculate the enthalpy change during a simple chemical reaction.

End of topic questions

1. A strip of magnesium ribbon 0.2 g is added to 40 cm³ of hydrochloric acid in a polystyrene beaker. The temperature rises by 32 °C. (The specific heat capacity of the hydrochloric acid can be assumed to be the same as that of water, that is, 4.2 J/g /°C.) Calculate:

   a) The energy released in the reaction. **(3 marks)**

   b) The energy released per gram of magnesium. **(1 mark)**

   c) The energy released per mole of magnesium. (A_r: Mg = 24) **(2 marks)**

2. An estimate of the energy produced when a fuel burns can be made by burning the fuel under a container holding water and measuring the temperature rise of the water.

   a) What type of material should the container be made of? Explain your answer. **(2 marks)**

   b) Why does this method give an estimate rather than an accurate value? **(2 marks)**

   c) How can the accuracy of this method be improved? **(2 marks)**

3. Calcium oxide reacts with water as shown in the equation:

   $$CaO(s) + H_2O(l) \rightarrow Ca(OH)_2(s)$$

   An energy level diagram for this reaction is shown below.

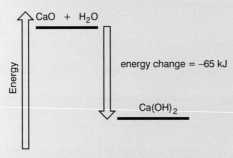

   a) What does the energy level diagram tell us about the type of energy change that takes place in this reaction? **(1 mark)**

   b) What does the energy level diagram indicate about the amounts of energy required to break bonds and form new bonds in this reaction? **(1 mark)**

4. Chlorine (Cl_2) and hydrogen (H_2) react together to make hydrogen chloride (HCl). The equation can be written as:

$$H–H + Cl–Cl \rightarrow H–Cl + H–Cl$$

When this reaction occurs, energy is transferred to the surroundings. Explain this in terms of the energy transfer processes taking place when bonds are broken and when bonds are made. **(2 marks)**

5. a) Calculate the energy change for the combustion of methane (CH_4) in oxygen (O_2). **(3 marks)**

$$+ 2 \times O{=}O \rightarrow O{=}C{=}O + 2 \times H–O–H$$

Use the bond energy values on page 205.

b) What does the sign of the energy change tell you about the reaction? **(1 mark)**

Rates of reaction

△ Fig. 3.10 Petrol igniting.

INTRODUCTION

Some chemical reactions take place extremely quickly. For example, when petrol is ignited it combines with oxygen almost instantaneously. Reactions like these have a *high rate*. Other reactions are much slower, for example when an iron bar rusts in the air; reactions like these have a *low rate*. Chemical reactions can be controlled and made to be quicker or slower. This can be very important in situations like food production, either by slowing down or increasing the rate at which food ripens, or in the chemical industry where the rate of the reaction can be adjusted to an optimum level.

KNOWLEDGE CHECK

✓ Know the arrangement, movement and energy of the particles in the three states of matter: solid, liquid and gas.
✓ Understand how the course of a reaction can be shown in an energy level diagram.
✓ Be able to write and interpret balanced chemical equations.

LEARNING OBJECTIVES

✓ Be able to describe experiments to investigate the effects of changes in surface area of a solid, concentration of a solution, temperature and the use of a catalyst on the rate of a reaction.
✓ Be able to describe the effects of changes in surface area of a solid, concentration of a solution, pressure of a gas, temperature and the use of a catalyst on the rate of a reaction.
✓ Be able to explain the effects of changes in surface area of a solid, concentration of a solution, pressure of a gas and temperature on the rate of a reaction in terms of particle collision theory.
✓ Know that a catalyst is a substance that increases the rate of a reaction, but is chemically unchanged at the end of the reaction.
✓ Know that a catalyst works by providing an alternative pathway with lower activation energy.
✓ Be able to draw and explain reaction profile diagrams showing ΔH and activation energy.
✓ Be able to investigate the effect of changing the surface area of marble chips and of changing the concentration of hydrochloric acid on the rate of reaction between marble chips and dilute hydrochloric acid.
✓ Be able to investigate the effect of different solids on the catalytic decomposition of hydrogen peroxide solution.

WHAT HAPPENS IN CHEMICAL REACTIONS?

A **chemical change**, or **chemical reaction**, is quite different from the **physical changes** that occur, for example, when sugar dissolves in water.

One or more new substances are produced.

In many cases an observable change is apparent, for example the colour changes or a gas is produced.

An apparent change in mass can occur. This change is often quite small and difficult to detect unless accurate balances are used. Mass is conserved in a chemical reaction – the apparent change in mass usually occurs because one of the reactants or products is a gas.

An energy change is almost always involved. In most cases energy is released and the surroundings become warmer. In some cases energy is absorbed from the surroundings and so the surroundings become colder. Note: Some physical changes, such as evaporation, also produce energy changes.

COLLISION THEORY

For a chemical reaction to occur, the reacting particles (atoms, molecules or ions) must collide. The energy involved in the collision must be enough to break the chemical bonds in the reacting particles – or the particles will just bounce off one another. This is the **collision theory**.

A collision that has enough energy to result in a chemical reaction is an **effective collision**.

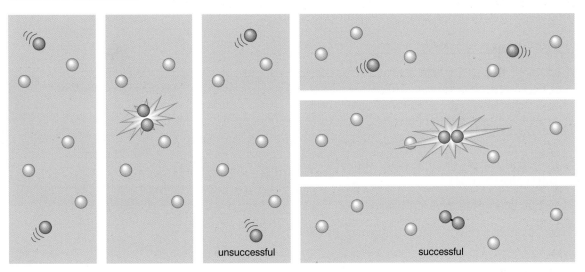

Δ Fig. 3.11 Particles must collide with sufficient energy to make an effective collision.

Some chemical reactions occur extremely quickly (for example, the explosive reaction between petrol and oxygen in a car engine) and some more slowly (for example, iron rusts over days or weeks). This is because they have different **activation energies**. Activation energy acts as a barrier to a reaction. It is the minimum amount of energy required in a collision for a reaction to occur. As a general rule, the bigger the activation energy the slower the reaction will be at a particular temperature.

The energy profile diagram in Fig. 3.12 shows the energy change against the course of the reaction. Initially any collision must generate energy equal to or above the activation energy and then energy is released as the reaction takes place. In this example the reaction is exothermic as the products have a lower energy level than the reactants and so energy will be lost to the surroundings. In this case, therefore, ΔH is negative indicating an exothermic reaction.

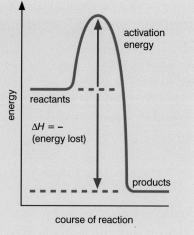

△ Fig. 3.12 Reaction profile.

REMEMBER

The 'barrier' preventing a reaction from occurring is called the activation energy. If the activation energy of a reaction is low, more of the collisions per second will be effective and the reaction will proceed quickly. If the activation energy is high a smaller proportion of collisions per second will be effective and the reaction will be slow.

RATE OF A REACTION

A quick reaction takes place in a short time. It has a high **rate of reaction**. As the time taken for a reaction to be completed increases, the rate of the reaction decreases. In other words:

Speed	Rate	Time
quick or fast	high	short
slow	low	long

QUESTIONS

1. In the collision theory what two things must happen for two particles to react?

2. What is an effective collision?

3. What is the activation energy of a reaction?

MONITORING THE RATE OF A REACTION

The rate of a reaction changes as the reaction proceeds. There are some easy ways of monitoring this change.

When marble (calcium carbonate) reacts with hydrochloric acid, the following reaction starts straight away:

calcium carbonate + hydrochloric acid → calcium chloride + carbon dioxide + water

$CaCO_3(s)$ + $2HCl(aq)$ → $CaCl_2(aq)$ + $CO_2(g)$ + $H_2O(l)$

The reaction can be monitored as it proceeds either by measuring the volume of gas being formed, or by measuring the change in mass of the reaction flask.

The volume of gas produced in this reaction can be measured using the apparatus shown in Fig. 3.13. The hydrochloric acid is put into the conical flask, the marble chips are added, the bung is quickly fixed into the neck of the flask and the stop clock is started.

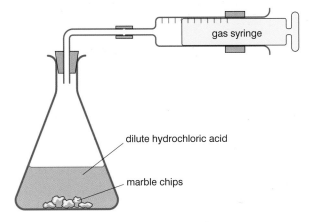

▷ Fig. 3.13 Monitoring the rate of a reaction.

The reaction will start immediately. Effervescence (bubbling) will occur in the flask as the carbon dioxide gas is produced and the plunger on the syringe will start to move. Measuring the volume of gas in the syringe every 10 seconds will show how the total amount of gas produced changes as the reaction proceeds. The change in the rate of the reaction with time can be shown on a graph of the results.

To measure the change in mass in the same reaction, the apparatus shown in Fig. 3.14 can be used. The hydrochloric acid is put into the conical flask, the marble chips are added, the cotton wool plug is put in the top of the flask and the stop clock is started. The mass of the flask and contents are measured as soon as the plug is inserted and then every 10 seconds as the reaction occurs. The mass will decrease as the carbon dioxide gas escapes from the flask.

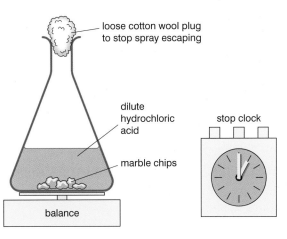

▷ Fig. 3.14 Measuring the change in mass.

As before, drawing a graph of the results shows the change in the rate of the reaction over time.

Graphs of the results from both experiments have almost identical shapes. The rate of the reaction decreases as the reaction proceeds.

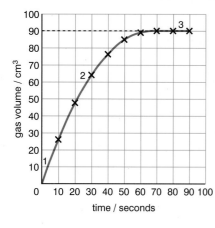

△ Fig. 3.15 Volume of carbon dioxide produced in reaction.

△ Loss in mass during the reaction.

The rate of the reaction at any point can be calculated from the gradient of the curve. The shapes of the graphs can be divided into three regions:

1. At this point, the curve is the steepest (has the greatest gradient) and the reaction has its highest rate. The maximum number of reacting particles are present and the number of effective collisions per second is at its greatest.

2. The curve is not as steep (has a lower gradient) at this point and the rate of the reaction is lower. Fewer reacting particles are present and so the number of effective collisions per second will be less.

3. The curve is horizontal (gradient is zero) and the reaction is complete. At least one of the reactants has been completely used up and so no further collisions can occur between the two reactants.

QUESTIONS

1. What piece of apparatus can accurately measure the volume of gas produced in a reaction?

2. On a volume versus time graph, what does a horizontal line show?

3. When comparing two reactions will the slower or quicker reaction have a steeper volume/ time gradient at the beginning?

WHAT CAN CHANGE THE RATE OF A REACTION?

There are five key factors that can change the rate of a reaction:

- concentration (of a solution)
- temperature
- surface area (of a solid)
- a catalyst
- pressure of a gas.

A simple collision theory can be used to explain how these factors affect the rate of a reaction. Two important parts of the theory are:

1. The reacting particles must collide with each other.

2. There must be sufficient energy in the collision to overcome the activation energy.

Concentration

Increasing the concentration of a reactant will increase the rate of a reaction. When a piece of magnesium ribbon is added to a solution of hydrochloric acid, the following reaction occurs:

magnesium	+	hydrochloric acid	$\rightarrow$	magnesium chloride	+	hydrogen
$Mg(s)$	+	$2HCl(aq)$	$\rightarrow$	$MgCl_2(aq)$	+	$H_2(g)$

As the magnesium and acid come into contact, there is effervescence ('fizzing'), and hydrogen gas is given off. Two experiments were performed using the same length of magnesium ribbon, but different concentrations of acid. In experiment 1 the hydrochloric acid used was 2.0 mol/dm³, in experiment 2 the acid was 0.5 mol/dm³. The graph shows the results of the two experiments.

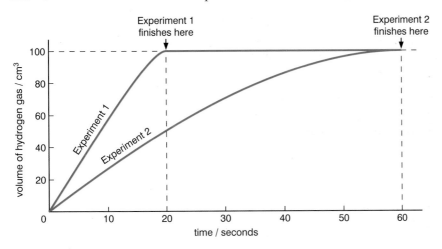

△ Fig. 3.16 Volume of hydrogen produced in the reaction between magnesium and hydrochloric acid.

In experiment 1 the curve is steeper (has a greater gradient) than in experiment 2. In experiment 1 the reaction is complete after 20 seconds, whereas in experiment 2 it takes 60 seconds. The rate of the reaction is higher with 2.0 mol/dm³ hydrochloric acid than with 0.5 mol/dm³ hydrochloric acid. In the 2.0 mol/dm³ hydrochloric acid solution the hydrogen ions are more likely to collide with the surface of the magnesium ribbon than in the 0.5 mol/dm³ hydrochloric acid.

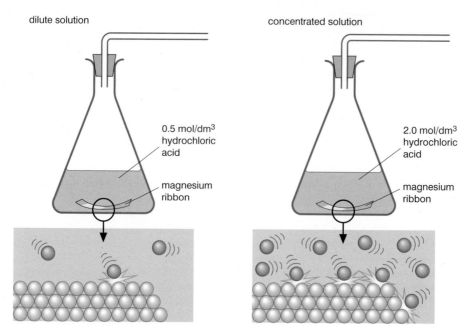

△ Fig. 3.17 Using dilute and concentrated solutions in a reaction.

Note: The same practical approach can be used to show the effect of changing the concentration of hydrochloric acid on the rate of reaction between marble chips and dilute hydrochloric acid.

Developing investigative skills

A student selected three samples of marble chips. Sample A contained large lumps, sample B contained smaller (middle-sized) lumps and sample C very small lumps (almost like a powder). The student set up some apparatus which could then be used to add 5 g of each sample to 50 cm³ of hydrochloric acid and measure the volume of gas produced. For each of the marble chip samples the student used a fresh sample of dilute hydrochloric acid.

The student recorded the results and then drew the graph shown below:

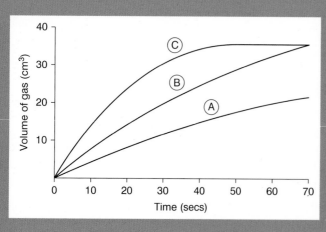

Devise and plan investigations

❶ What piece of equipment did the student use to measure accurately the volume of gas produced in each experiment?

❷ Suggest how the student could have added the samples of marble chip to the acid without losing some of the gas produced?

Analyse and interpret

❸ Reaction A had not finished after 70 seconds. How can you deduce this from the graph of the results?

❹ For reaction C at what point was the reaction rate greatest?

Evaluate data and methods

❺ What is the name of the gas produced in the reactions?

❻ Explain why reaction C had the greatest initial rate.

❼ Explain why in reaction C no more gas is produced after 40 seconds.

Developing investigative skills

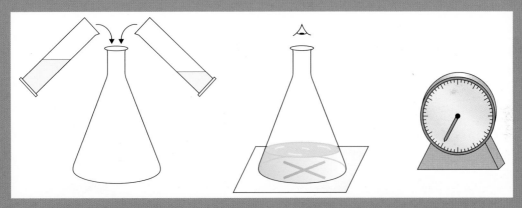

△ Fig. 3.18 Experiment with sodium thiosulfate solution and hydrochloric acid.

A student was investigating the reaction between sodium thiosulfate solution and dilute hydrochloric acid. As the reaction takes place a precipitate of sulfur forms in the solution and makes the solution change from colourless (and clear) to pale yellow (and opaque). The time it takes for a certain amount of sulfur to form can be used as a measure of the rate of the reaction. (Note: sulfur dioxide is produced in this reaction and the gas is toxic and asthmatics are particularly at risk; dilute sulfuric acid may cause harm in eyes or a cut.)

❶ The student used 1.0 mol/dm³ sodium thiosulfate solution and made up different concentrations of the solution by using the quantities of the solution and water shown in the table.

❷ The student then drew a mark in pencil on a piece of paper.

❸ The student then added 5 cm³ of dilute hydrochloric acid to the solution in one of the flasks, the clock was started, the mixture was quickly stirred or swirled and then the conical flask was put on top of the pencilled cross.

❹ The student looked down through the conical flask to the mark and stopped the clock as soon the mark could no longer be seen.

❺ The student then repeated the process with the other four solutions. The results are shown in the table below:

Volume of sodium thiosulfate solution (cm³)	50	40	30	20	10
Volume of water (cm³)	0	10	20	30	40
Volume of hydrochloric acid (cm³)	5	5	5	5	5
Time for the mark to be obscured (s)	14	18	23	36	67

△ Table 3.4 Results of experiment.

Devise and plan investigations

❶ Why was the total volume in the flask always 55 cm³?

❷ Why was the clock started when the acid was added and not when the flask was put on the pencilled mark?

❸ What apparatus would you use to measure the volume of sodium thiosulfate solution?

Analyse and interpret

Draw a graph of volume of sodium thiosulfate solution against time. Draw a smooth curve through the points.

❹ What does the overall shape of the curve tell you about the effect of changing the volume of sodium thiosulfate on the rate of the reaction?

❺ Use your results to predict what the time would have been if 15 cm³ of sodium thiosulfate and 35 cm³ of water had been used.

Evaluate data and methods

❻ What do you think are the main sources of error in this experiment? (Pick the two which you think would have the greatest effect on the accuracy of the results.)

Temperature

Increasing the temperature increases the rate of reaction. Warming a chemical transfers kinetic energy to the chemical's particles. More kinetic energy means that the particles move faster. As they are moving faster there will be more collisions each second. The increased energy of the collisions also means that the proportion of collisions that are effective will increase.

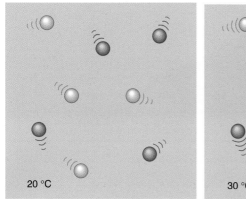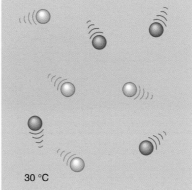

△ Fig. 3.19 Effect of increasing temperature on particles.

Increasing the temperature of a reaction such as that between calcium carbonate and hydrochloric acid will not increase the final amount of carbon dioxide produced. The same amount of gas will be produced in a shorter time. The rates of the two reactions are different but the final loss in mass is the same.

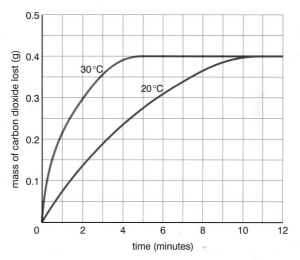

△ Fig. 3.20 The effect of temperature on the reaction between hydrochloric acid and calcium carbonate.

QUESTIONS

1. What units are used to measure the concentration of solutions?

2. In terms of particles colliding, why does increasing the concentration of solution increase the rate of reaction?

3. Give two reasons why increasing temperature increases the rate of reaction.

Surface area

Increasing the **surface area** of a solid reactant will increase the rate of a reaction. The reaction can only take place if the reacting particles collide. This means that the reaction takes place at the surface of the solid. The particles within the solid cannot react until those on the surface have reacted and moved away.

Powdered calcium carbonate has a much larger surface area than the same mass of marble chips. A lump of coal will burn slowly in the air, whereas coal dust can react explosively.

Δ Fig. 3.21 Powdered carbon has a much larger surface area than the same mass in larger lumps.

SCIENCE IN CONTEXT

THE EXPLOSIVE TRUTH ABOUT FLOUR MILLS

The surface area of particles really does affect the rate of some reactions!

Baking bread is a common and important activity but making the flour that goes into the bread can be a dangerous business. A serious explosion at a flour mill near Minneapolis in the USA in 1878 killed 18 people. Ever since then, the milling industry has tried to reduce the risk of flour particles igniting into 'flour bombs'. In fact flour dust is thought to be more explosive than coal dust! Similar explosions have occurred in other factories when dust itself has exploded.

Δ Fig. 3.22 Dropping milk powder on a flame.

The key components to a flour or dust explosion are very small particles suspended in a plentiful supply of air, in a confined space and with a source of ignition. In factories the source of ignition doesn't have to be something obvious such as a discarded cigarette or match. It could be a spark from an electrical motor or other electrical device, even a light switch. In the case of the 1878 flour mill explosion the cause of the explosion was thought to be a spark from an old electric motor.

In the laboratory, if you put a match to some flour you might be able to get it to burn but it certainly won't explode – the flour needs to be suspended in the air as very small particles which are close enough together so that if one flour particle ignites it forms a rapid chain reaction with all the other particles and then an explosion. So don't forget the importance of surface area on the rate of some reactions.

Catalysts

A catalyst is a substance that increases the rate of a chemical reaction and is chemically unchanged at the end of the reaction. An enzyme is a biological catalyst, for example amylase which is found in saliva.

Most catalysts work by providing an alternative 'route' for the reaction which has a lower activation energy. The lower activation energy means that more of the collisions per second between particles will be effective.

An example of the effect of a catalyst on a reaction is the use of manganese(IV) oxide in the decomposition of hydrogen peroxide. Hydrogen peroxide decomposes at room temperature into water and oxygen. The rate of this reaction is considerably increased by adding manganese(IV) oxide. As a gas is produced the rate of the reaction can be monitored by collecting the gas in a gas syringe.

△ Fig. 3.23 The effect of a manganese(IV) oxide catalyst on the decomposition of hydrogen peroxide.

$$2H_2O_2 \, (aq) \rightarrow 2H_2O \, (l) + O_2 \, (g)$$

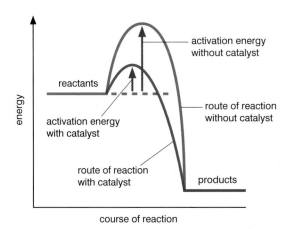

△ Fig. 3.24 The catalyst provides a lower energy route from reactants to products.

Catalysts are often used in industry to manufacture important chemicals. Table 3.5 includes some important industrial catalysts.

Industrial process	Catalyst used
Manufacture of ethanol	Phosphoric acid
Cracking long-chain alkanes	Silica or alumina
Manufacture of ammonia	Iron
Manufacture of sulfuric acid	Vanadium(V) oxide

△ Table 3.5 Uses of catalysts.

REMEMBER

Enzymes are often very specific to a particular reaction. They have an 'active site' which is just the right shape for the reacting particles to fit into. Molecules with other structures and shapes cannot do this. Metals, such as iron used in the manufacture of ammonia, work in the same sort of way that enzymes do. The surface of the iron allows molecules of nitrogen and hydrogen to get 'trapped'. They then collide more frequently in the confined space and an effective collision becomes more likely.

Developing investigative skills

A student was set the challenge of investigating which of three different solids was the most effective catalyst for decomposing hydrogen peroxide (H_2O_2) into water and oxygen gas. The student decided to make sure all the solids were in powder form. The student then set up an apparatus involving a conical flask, a small test tube with cotton thread attached to its rim, and a delivery tube connected to a gas syringe. The student put $50\,cm^3$ of hydrogen peroxide solution into the conical flask and carefully suspended the small test tube containing solid A in the solution. When all the apparatus was connected the student carefully tipped the flask until the solution and the solid came into contact. The student then started the stop clock and measured the amount of oxygen produced every 10 seconds for 80 seconds. The results obtained are shown in the table.

Solid		Time (secs)								
		0	10	20	30	40	50	60	70	80
A	volume oxygen (cm^3)	0	15	28	39	44	54	57	57	57
B	volume oxygen (cm^3)	0	10	18	24	27	28	28	28	28
C	volume oxygen (cm^3)	0	0	0	0	0	0	0	0	0

Demonstrate and describe techniques

❶ Draw and label a diagram to show the apparatus used to measure the volume of oxygen produced.

❷ Why do you think the student decided to make sure all the solids were powders rather than some being powders and others lumps?

Analyse and interpret data

❸ Write a fully balanced symbol equation with state symbols for the reaction.

❹ Draw a graph of the results showing the three set of results on the same graph.

❺ Which solid was the most effective catalyst?

❻ For A and B the curves you have drawn should be steepest between 0 and 10 seconds. Give an explanation for this.

❼ Suggest a reason why for A and B no more gas is produced towards the end of the 80 second period.

Evaluate data and methods

❽ Do you think putting the solids in small test tubes suspended in the solution helped to produce accurate results? Explain your answer.

❾ One of the results obtained is anomalous. What does this mean and which result do you think it is?

Pressure

Changing the pressure can alter the rate of a chemical reaction, particularly when the reaction is between gases. Increasing the pressure of the reaction between gases will increase the rate of the reaction. Increasing the pressure has the effect of reducing the volume of the gas and so moving the particles closer together. As the particles are closer together there will be more collisions in a certain time and therefore more effective collisions.

low pressure

high pressure

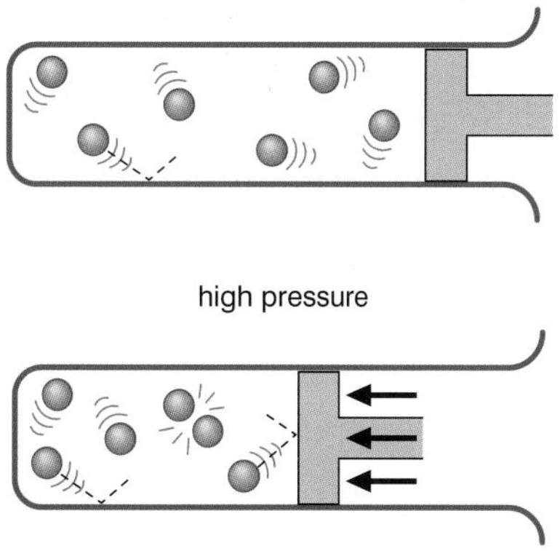

△ Fig. 3.25 The same number of particles are closer together in a smaller volume. There will be more effective collisions each second.

QUESTIONS

1. What is a catalyst?

2. What is a biological catalyst usually called?

3. In what sort of reactions will changes in pressure cause changes in the rate of reaction?

EXTENSION

Time is money in a competitive world, so it is important to understand what affects the rate of a chemical reaction. The rate can be measured by measuring the rate of formation of the product or the rate of removal of a reactant. Use your practical knowledge and scientific understanding to answer the questions below.

1. A group of students studying rate of reaction were given several strips of magnesium ribbon and dilute hydrochloric acid.

   a) Write a word and balanced symbol equation for this reaction. (Hydrogen gas (H_2) and magnesium chloride ($MgCl_2$) are the products formed.)

   b) They carried out two experiments to check the effect of temperature on the rate of the reaction. A lower temperature was used in the second experiment. In the first experiment the students measured the volume of hydrogen given off over a period of time.

   i)   Draw a labelled diagram to show the apparatus that the students could use.

   ii) Indicate what must be kept constant in both experiments so the effect of temperature on the rate of reaction can be compared.

   iii) Draw two labelled sketch graphs showing the change of volume of gas during each of the two experiments.

2. Using the sketch graphs, show how the initial speed of each of the reactions can be found and comment how these values should compare.

End of topic checklist

An **effective collision** is one that has enough energy to cause a chemical reaction.

Activation energy is the minimum energy that must be provided before a reaction will take place.

A **catalyst** is a substance that increases the rate of a chemical reaction and is chemically unchanged at the end of the reaction.

The facts and ideas that you should understand by studying this topic:

○ Know the relationship between the speed and rate of a reaction.

○ Be able to describe experiments to investigate how the rate of a reaction can be affected by:

- Changing the concentration of a solution;
- Changing the temperature of the reaction;
- Changing the surface area of a solid;
- Using a catalyst;
- Changing the pressure of reacting gases.

○ Be able to investigate the effect of changing the surface area of marble chips and of changing the concentration of hydrochloric acid on the rate of reaction between marble chips and dilute hydrochloric acid.

○ Be able to investigate the effect of different solids on the catalytic decomposition of hydrogen peroxide solution.

○ Know how a gas syringe or a balance can be used to measure the rate of a reaction in which a gas is produced.

○ Know that a collision between particles must take place and there must be enough energy in the collision to break chemical bonds before a reaction can take place.

○ Understand that the energy barrier that must be overcome before a reaction can take place is called the activation energy.

○ Be able to draw and explain reaction profile diagrams showing ΔH and activation energy.

○ Understand how increasing the concentration of a solution will increase the rate of a reaction involving that solution by increasing the number of effective collisions.

○ Understand how increasing the temperature of a reaction will increase the rate of that reaction by increasing the number of effective collisions.

○ Understand how increasing the surface area of a solid will increase the rate of a reaction involving that solid by increasing the number of effective collisions.

○ Understand how a catalyst can increase the rate of a reaction by providing an alternative route with a lower activation energy.

○ Understand how increasing the pressure applied to gases when they react will increase the rate of reaction by increasing the number of effective collisions per second.

End of topic questions

1. For a chemical reaction to occur the reacting particles must collide. Why don't all collisions between the particles of the reactants lead to a chemical reaction?

(2 marks)

2. The diagrams show the activation energies of two different reactions, A and B.

a) What is the activation energy of a reaction? (1 mark)

b) Which reaction is likely to have the greater rate of reaction at a particular temperature?

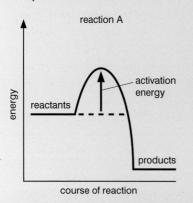
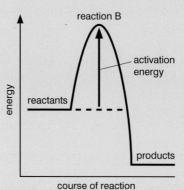

Explain your answer. (2 marks)

3. Look at the table of results obtained when dilute hydrochloric acid is added to marble chips. (Marble chips are in excess.)

Time (seconds)	0	10	20	30	40	50	60	70	80	90
Volume of gas (cm³)	0	20	36	49	58	65	69	70	70	70

a) What is the name of the gas produced in this reaction? (1 mark)

b) Write a balanced symbol equation, including state symbols, for the reaction.

(2 marks)

c) Draw a graph of volume of gas (*y*-axis) against time (*x*-axis). Label the graph 'graph 1'. (3 marks)

d) Use the results to calculate the volume of gas produced:

   i) in the first 10 seconds (1 mark)

   ii) between 10 and 20 seconds (1 mark)

   iii) between 20 and 30 seconds (1 mark)

   iv) between 80 and 90 seconds. (1 mark)

e) Explain how the rate of the reaction changes as the reaction takes place.

(2 marks)

f) Use collision theory to explain the change in the rate of reaction. (2 marks)

g) The reaction was repeated using the same volume of hydrochloric acid and the same mass of marble but as a powder instead of chips. Draw another curve on your graph paper, using the same axes as before (label as graph 2), to show how the original results will change. (3 marks)

h) The reaction was repeated again but this time using the original mass of new marble chips and the same volume of hydrochloric acid, but with the acid only half as concentrated as originally. Draw another curve on your graph paper, using the same axes as before (label graph 3), to show how the original results will change. (3 marks)

4. Explain why increasing the temperature of a reaction will increase the rate of the reaction. (2 marks)

5. a) Explain how a catalyst increases the rate of a reaction. (2 marks)

b) Name a catalyst that will increase the rate of decomposition of hydrogen peroxide. (1 mark)

c) Name the catalyst used in the industrial manufacture of ethanol from ethene. (1 mark)

6. The equation for the manufacture of ammonia from nitrogen and hydrogen is shown below:
$$N_2(g) + 3H_2(g) \rightarrow 2NH_3(g)$$
Would you expect the rate of this reaction to be affected by increasing the pressure? Explain your answer. (2 marks)

△ Fig. 3.26 When copper(II) sulfate crystals are heated they turn from blue to white. The reaction can then be reversed by adding water.

Reversible reactions and equilibria

INTRODUCTION

Many of the reactions used in the manufacture of important chemicals 'go both ways'. In other words, the reactants form the products, but they revert to the reactants – sometimes at the same time as they form them. These reversible reactions came as something of a shock to early chemists, but today chemists have found ways to maximise the direction of the reaction and so produce as much product as possible. If all the products and reactants are kept in a closed system (that is, nothing is allowed to escape), all reversible reactions reach a 'balance point' or equilibrium. This balance point may be at a point where the concentrations of reactants and products are the same, but this is rarely the case – the balance point often has either more reactants or more products. For industrial chemists, altering the balance point or equilibrium is a key part of their work.

KNOWLEDGE CHECK

✓ Understand what is meant by the rate of a reaction.
✓ Know what enthalpy changes are associated with exothermic and endothermic reactions.
✓ Be able to interpret chemical equations and associated state symbols.

LEARNING OBJECTIVES

✓ Know that some reactions are reversible and this is indicated by the symbol $\rightleftharpoons$ in equations.
✓ Describe reversible reactions such as the dehydration of hydrated copper(II) sulfate and the effect of heat on ammonium chloride.
✓ Know that a reversible reaction can reach dynamic equilibrium in a sealed container.
✓ Know that the characteristics of a reaction at dynamic equilibrium are:
  • the forward and reverse reactions occur at the same rate
  • the concentrations of reactants and products remain constant.
✓ Understand why a catalyst does not affect the position of equilibrium in a reversible reaction.
✓ Know the effect of changing either temperature or pressure on the position of equilibrium in a reversible reaction:
  • an increase (or decrease) in temperature shifts the position of equilibrium in the direction of the endothermic (or exothermic) reaction
  • an increase (or decrease) in pressure shifts the position of equilibrium in the direction that produces fewer (or more) moles of gas.

TYPES OF REVERSIBLE REACTION

Carbon burns in oxygen to form carbon dioxide:

carbon + oxygen → carbon dioxide

$C(s)$ + $O_2(g)$ → $CO_2(g)$

Carbon dioxide cannot easily be changed back into carbon and oxygen. The reaction cannot be easily reversed.

When blue copper(II) sulfate crystals are heated, a white powder is formed (anhydrous copper(II) sulfate) and water is lost as steam. If water is added to this white powder, blue copper(II) sulfate is re-formed. The reaction is **reversible**:

copper(II) sulfate crystals $\rightleftharpoons$ anhydrous copper(II) sulfate + water

$CuSO_4 \cdot 5H_2O(s)$ $\rightleftharpoons$ $CuSO_4(s)$ + $5H_2O(l)$

A reversible reaction can go from left to right or from right to left. Notice the double-headed arrow $\rightleftharpoons$ used when writing these equations.

Another example of a reversible reaction is the decomposition of ammonium chloride. If ammonium chloride is heated in a long tube, the solid produces colourless fumes of a mixture of gases. In the cooler parts of the tube the fumes reform the white solid. The equation is:

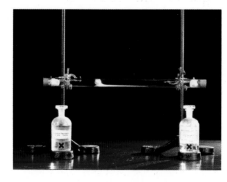

ammonium chloride $\rightleftharpoons$ ammonia + hydrogen chloride

$NH_4Cl(s)$ $\rightleftharpoons$ $NH_3(g)$ + $HCl(g)$

△ Fig. 3.27 Hydrochloric acid and ammonia react to form ammonium chloride. When heated, ammonium chloride forms ammonia and hydrogen chloride.

The reaction between ethene and water to make ethanol is also a reversible reaction. This is one of the reactions used industrially to make ethanol:

ethene + water $\rightleftharpoons$ ethanol

$C_2H_4(g)$ + $H_2O(g)$ $\rightleftharpoons$ $C_2H_5OH(g)$

When ethene and water are heated in the presence of a catalyst in a sealed container, ethanol is produced.

As the ethene and water are used up, the rate of the forward reaction decreases. As the amount of ethanol increases, the rate of the back reaction (the decomposition of ethanol) increases. Eventually the rate of formation of ethanol will exactly equal the rate of decomposition of ethanol. The amounts of ethene, water and ethanol will be constant. The reaction is said to be in **equilibrium**.

Dynamic equilibrium

A chemical equilibrium is an example of a dynamic equilibrium (a moving equilibrium). In a sealed container, reactants are constantly forming products, and products are constantly reforming the reactants. Equilibrium is reached when the rates of the forward and backward reactions are the same. At equilibrium the concentrations of products and reactants remain constant.

To get an idea of a dynamic or moving equilibrium, imagine you are walking up an escalator as the escalator is moving down. You are still moving forward and the escalator is moving towards you. If your rate of movement is the same as the escalator's rate of movement, but in the opposite direction, you will not appear to be moving. This 'moving balance point' is called a dynamic equilibrium. If you imagine a situation where you start to walk up the escalator when it is moving slowly in the opposite direction, but gradually speeds up to the speed you are walking, you might get almost to the top of the escalator before your rates of movement are balanced. Alternatively, if the escalator starts slowly but then very quickly reaches your rate of movement, the balance point may be near the bottom of the escalator. In the same sort of way, chemical reactions can have very different balance points (nearer to the reactants or nearer to the products).

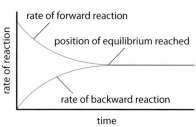

△ Fig. 3.28 At equilibrium the rate of the forward reaction equals the rate of the backward reaction.

QUESTIONS

1. Use the example of copper(II) sulfate to explain what a reversible reaction is.

2. What does it mean if a reaction is in equilibrium?

3. An equilibrium in a chemical reaction is dynamic. What does this mean?

CHANGING THE POSITION OF EQUILIBRIUM

Reversible reactions can be a nuisance to an industrial chemist. You want to make a particular product, but as soon as it forms it starts to change back into the reactants! Fortunately, scientists have found ways of increasing the amount of product that can be obtained (the yield) in a reversible reaction by moving the position of balance to favour the products rather than the reactants.

The position of equilibrium or yield can be changed in the following ways:

- changing concentrations
- changing pressure
- changing temperature.

To be able to predict how the position of the equilibrium will change, it is useful to remember that whatever change is applied to the reaction, the reaction will try to reduce the effect of the change. Reactions are awkward!

In the following example:

$A(g) + 2B(g) \rightleftharpoons 2C(g)$ ΔH positive

Change made	Effect on the equilibrium position	Method of predicting the effect
Increasing the concentration of A or B	Moves to the right-hand side	The equilibrium moves in a direction that reduces the concentration of A or B. It does this by converting A and B into C.
Decreasing the concentration of C	Moves to the right-hand side	The equilibrium moves in a direction that increases the concentration of C. It does this by converting A and B into C.
Increasing the pressure acting on the reaction	Moves to the right-hand side	The equilibrium moves in the direction that produces the fewer molecules/moles of gas. There are fewer molecules/moles of gas on the right-hand side. It does this by converting A and B into C.
Increasing the temperature of the reaction	Moves to the right-hand side	The equilibrium moves in the direction that absorbs heat energy, i.e. the endothermic reaction. The forward reaction is endothermic (ΔH is positive) so A and B are converted into C.

△ Table 3.6 Effects of changes on the equilibrium position.

Note: If the forward reaction is endothermic (ΔH positive) then the backward reaction will be exothermic (ΔH negative).

A catalyst increases the rate at which the equilibrium is achieved. As it changes the rates of both the forward and reverse reactions, it does not change the position of the equilibrium. It does not change the yield either.

EQUILIBRIA – IT'S CERTAINLY A BALANCING ACT!

Ammonia is an important chemical used in the manufacture of nitric acid and fertilisers. It is made from nitrogen and hydrogen as shown in the equation:

$$N_2(g) + 3H_2(g) \rightleftharpoons 2NH_3(g) \qquad \Delta H = \text{exothermic}$$

The reaction is reversible and the forward reaction is exothermic. This causes a problem in the industrial manufacture of ammonia. In terms of the equilibrium, and getting the maximum conversion of nitrogen and hydrogen into ammonia, the temperature of the reaction needs to be as low as possible. However, when the temperature is low the rate of the reaction is very low. In other words, while a low temperature would give a good conversion into ammonia, it would take a very long time to achieve this.

△ Fig. 3.29 Ammonium nitrate fertiliser is made from ammonia and nitric acid.

Chemists are used to balancing the opposing benefits of *rate* and *equilibrium* in reactions like these. They have to find an acceptable compromise. These are the solutions:

1. A catalyst (iron) is used to increase the rate of the reaction. This does not affect the position of equilibrium at all.

2. A high pressure (200 atmospheres) is used. It affects the equilibrium position and produces a higher proportion of ammonia. (There are four moles of reactant molecules and only two moles of product molecules – increasing the pressure favours conversion to the smaller number of molecules, that is, the ammonia.) Note that using a higher pressure would improve conversion to ammonia, but ordinary reaction vessels would not be strong enough to cope with the higher pressure, and strong-enough vessels would be too expensive to produce.

3. A temperature of 450 °C is used. The graph helps to explain this choice:

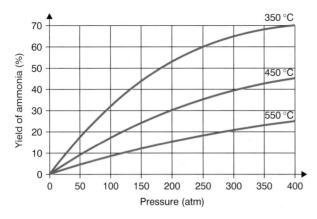

△ Fig. 3.30 Effect of temperature and pressure on yield of ammonia

You can see from the graph that a lower temperature (350 °C) gives a better conversion to ammonia but this would be a slower reaction. A higher temperature (550 °C) would give a better rate of reaction but a poorer conversion to ammonia. The compromise temperature is 450 °C.

The conditions have been skilfully manipulated to produce the best yield at the right cost.

QUESTIONS

1. This question is about the reaction of ethene and steam:

$$C_2H_4(g) + H_2O(g) \rightleftharpoons C_2H_5OH(g) \qquad \Delta H = \text{negative}$$

Which conditions would maximise the amount of ethanol in the equilibrium mixture?

a) High or low temperature

b) High or low pressure

c) Catalyst or no catalyst

2. Explain why a catalyst does not affect the position of equilibrium in a reversible reaction.

End of topic checklist

A **reversible reaction** is one in which reactants form products and products can reform reactants.

In a **dynamic equilibrium** reactants are constantly being converted into products and products are constantly being converted back into reactants. The rates of the forward and backward reactions are the same and the concentration of reactants and products stay the same.

The facts and ideas that you should know and understand by studying this topic:

○ Understand that some reactions, such as heating copper(II) sulfate or ammonium chloride, are reversible.

○ Know the symbol used in a chemical equation to show a reversible reaction.

○ Be able to describe the changes that occur when hydrated copper(II) sulfate is heated and then when water is re-added.

○ Be able to describe the changes that occur when ammonium chloride is heated.

○ Understand the concept of a dynamic equilibrium.

○ Be able to predict the effects of the following changes on a reversible or equilibrium reaction:

- changing the concentration of reactants or products;
- changing the pressure;
- changing the temperature on both endothermic and exothermic reactions.

○ Know that a catalyst affects the rates of forward and backward reactions to the same extent and so does not change the position of equilibrium.

End of topic questions

1. What is a reversible reaction? **(1 mark)**

2. What does it mean when a chemical reaction is in dynamic equilibrium? **(2 marks)**

3. Describe what you would observe in the following reactions:

   a) Hydrated copper(II) sulfate is heated strongly until there is no further change.
   (3 marks)

   b) After cooling, water is added to the product formed in reaction **a)**. **(2 marks)**

4. Write a balanced symbol equation, including state symbols, for the reaction referred to in question **3** above. **(3 marks)**

5. Some ammonium chloride is heated strongly in a boiling tube. Describe what you would expect to observe. **(3 marks)**

6. The reaction between sulfur dioxide and oxygen is reversible as shown by the equation:

   $$2SO_2(g) + O_2(g) \rightleftharpoons 2SO_3(g) \qquad \Delta H \text{ negative}$$

   What effect will the following changes have on the yield of sulfur trioxide (SO_3)?

   a) increasing the volume of oxygen. Explain your answer. **(2 marks)**

   b) increasing the pressure on the reaction. Explain your answer. **(2 marks)**

   c) increasing the temperature of the reaction. Explain your answer. **(2 marks)**

7. What effect does a catalyst have on:

   a) the rate of reaction? Explain your answer. **(2 marks)**

   b) the position of equilibrium or yield of a reaction? Explain your answer. **(2 marks)**

Exam-style questions
Sample student answers

Question 1

The following equation represents what happens when sulfur dioxide (SO_2) reacts with oxygen (O_2) to form sulfur trioxide (SO_3):

$$2SO_2(g) + O_2(g) \rightarrow SO_3(g)$$

In both SO_2 and SO_3 each oxygen atom forms double bonds to the sulfur atom. The two oxygen atoms in the O_2 molecule are joined with a double bond.

The table below shows the mean bond energies.

Bond	Mean bond energy (kJ/mol)
O=O	496
S=O	493

a) Which bonds are broken in the reaction? (2)

4 O=S bonds and 1 O=O bond ✓ ②

b) Which bonds are formed in the reaction? (2)

6 O=S bonds ✓ ②

c) Use the mean bond energies provided to calculate the enthalpy change (ΔH) for the reaction. (4)

Bonds broken = 4(493) + 496 = 2468 ✓

Bonds formed = 6(493) = 2958 ✓

Energy change = 490 ✓ (kJ) ③

EXAMINER'S COMMENTS

a) Correct. The bonds have been correctly identified and the number of each.

b) Correct. Again the bond has been correctly identified and the number of this type of bond.

c) The student has done the calculation correctly with the working clearly shown. 1 mark is lost as the unit is incorrect – should be kJ/mol.

d) Correct. It would have been sufficient to say that the enthalpy change was negative.

e) The energy diagram has been drawn correctly but a mark has been lost for not labelling the enthalpy change (−)490 kJ/mol.

Exam-style questions cont.

d) Is the reaction endothermic or exothermic? Explain your answer. (1)

> *More energy is released on forming bonds than was used in breaking bonds so the reaction is exothermic.* ✓

e) Draw an energy level diagram to represent the reaction. (2)

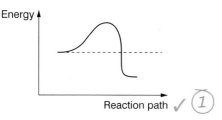

(Total 11 marks)

$\frac{9}{11}$

Question 2

Dilute nitric acid reacts with marble chips to produce carbon dioxide. The equation is given below:

$$2HNO_3(aq) + CaCO_3(s) \rightarrow Ca(NO_3)_2(aq) + H_2O(l) + CO_2(g)$$

Some students investigated the effect of changing the temperature of the nitric acid on the rate of the reaction. The method is:

- Use a measuring cylinder to pour $50\,cm^3$ of dilute nitric acid into a conical flask.
- Heat the acid to the required temperature.
- Place the flask on the balance.
- Add 15 g (an excess) of marble chips to the flask.
- Time how long it takes for the mass to decrease by 1.00 g.
- Repeat the experiment at different temperatures.

The students' results are shown in the table.

Temperature of acid (°C)	Time to lose 1.00 g (s)
20	93
33	68
44	66
55	40
67	30
76	25

a) i) Draw a graph of the results. .... **(3)**

ii) One of the points is inaccurate. Circle this point on your graph. .... **(1)**

iii) Suggest a possible cause for this inaccurate result. .... **(1)**

b) Use the graph to find the time taken to lose 1.00 g at 40 °C and 60 °C. .... **(2)**

c) i) The rate of the reaction can be found using the equation:

$$\text{rate of reaction} = \frac{\text{mass lost}}{\text{time taken to lose mass}}$$

Use this equation and your results from **b)** to calculate the rates of reaction at 40 °C and 60 °C. .... **(2)**

ii) What will be the unit for these rates? .... **(1)**

iii) State how the rate of reaction changes when the temperature increases. .... **(1)**

iv) Explain in terms of particles and collisions why the rate changes when the temperature increases. .... **(1)**

d) Describe how the method could be changed to obtain a result at 5 °C. .... **(3)**

(Total 15 marks)

Question 3

The equation for the reaction between nitrogen and oxygen to form nitrogen oxide is shown below:

$$N_2(g) + O_2(g) \rightleftharpoons 2NO(g) \; \Delta H \text{ positive}$$

What effect will the following changes have on the amount of nitrogen oxide in the equilibrium mixture?

a) Increasing the concentration of oxygen. Explain your answer. .... (2)

b) Increasing the pressure on the reaction. Explain your answer. .... (2)

c) Increasing the temperature of the reaction. Explain your answer. .... (2)

(Total 6 marks)

Question 4

Ammonia (NH_3) is manufactured from nitrogen (N_2) and hydrogen (H_2) in the presence of an iron catalyst. The reaction is reversible. Explain what effect the catalyst has on the conversion of nitrogen and hydrogen into ammonia in this reaction. .... (3)

(Total 3 marks)

Question 5

Use the average bond energies provided to calculate the molar enthalpy change (ΔH) in the following reaction:

$$CH_4(g) \;+\; 2O_2(g) \;\rightarrow\; CO_2(g) \;+\; 2H_2O(l)$$

(Average bond energies: C–H 413 kJ/mol; O=O 498 kJ/mol; C=O 745 kJ/mol; H–O 464 kJ/mol)

(Total 8 marks)

Question 6

2g of magnesium ribbon was added to 100 cm³ of 1 mol/dm³ hydrochloric acid in a polystyrene cup. The temperature rose from 18 °C to 30 °C.

$$Mg(s) \;+\; 2HCl(aq) \;\rightarrow\; MgCl_2(aq) \;+\; H_2(g)$$

a) Calculate the heat energy change in the reaction.

b) Calculate the molar enthalpy change in the reaction. (Note: you will need to decide which reagent is in excess).

(Relative atomic mass of Mg = 24; Assume the specific heat capacity of the acid is 4.2 J/g/°C)

(Total 6 marks)

Organic chemistry is one of the 'branches' of chemistry and is seen as distinct from other branches, such as inorganic and physical chemistry. It can be described as the chemistry of living processes (often referred to as biochemistry) but extends beyond that. It focuses almost entirely on the chemistry of covalently bonded carbon molecules and, as well as life processes, it includes the chemistry of other types of compounds, including plastics, petrochemicals, drugs and paint.

The early chemists didn't think they would ever be able to make the sort of chemicals involved in living processes but they were wrong. For example, today very complex chemicals used in the manufacture of drugs can be made and then their structures modified to achieve improvements in their effectiveness.

An understanding of organic chemistry can be developed from a knowledge of the structure of a carbon atom and how it can combine with other carbon atoms by forming covalent bonds. In this section you will be introduced to a few of the 'families' or series of organic compounds. This knowledge will provide a sound basis for further work in chemistry or biology.

STARTING POINTS

1. Where is carbon in the Periodic Table of elements? What can you work out about carbon from its position?

2. What is the atomic structure of carbon? How are its electrons arranged?

3. How does carbon form covalent bonds? Show the bonding in methane (CH_4), the simplest of organic molecules.

4. You will be learning about series of organic compounds which are hydrocarbons. What do you think a hydrocarbon is?

5. You will be learning about methane. Where can methane be found and what it is used for?

6. You will also be learning about ethanol, which belongs to a particular series of organic compounds. Do you know where you could find ethanol in everyday products?

SECTION CONTENTS

a) Crude oil

b) Alkanes

c) Alkenes

d) Alcohols

e) Carboxylic acids

f) Esters

g) Synthetic polymers

h) Exam-style questions

4

Organic chemistry

△ Many paints contain hydrocarbons, which are organic chemicals.

Crude oil

△ Fig. 4.1 Crude oil contains a mixture of hydrocarbons.

INTRODUCTION

Crude oil is a rich source of important chemicals. Many of them are familiar fossil fuels used in the home and for transport. Separating the components of crude oil is the first stage of refining oil. The second stage is to convert some of the less useful components into chemicals for industry: for example, many of the raw materials used in plastics come from crude oil. However, the world's supplies of crude oil are running out, and the search is on to find alternatives.

KNOWLEDGE CHECK

✓ Know that carbon compounds are covalently bonded.
✓ Know that the burning of fossil fuels produces carbon dioxide, a greenhouse gas.
✓ Know that there are alternative energy sources to fossil fuels.

LEARNING OBJECTIVES

✓ Know that a hydrocarbon is a compound of hydrogen and carbon only.
✓ Know that crude oil is a mixture of hydrocarbons.
✓ Know what is meant by the term homologous series.
✓ Be able to describe how the industrial process of fractional distillation separates crude oil into fractions.
✓ Know the names and uses of the main fractions obtained from crude oil: refinery gases, gasoline, kerosene, diesel, fuel oil and bitumen.
✓ Know the trend in colour, boiling point and viscosity of the main fractions.
✓ Know that a fuel is a substance that, when burned, releases heat energy.
✓ Know the possible products of complete and incomplete combustion of hydrocarbons with oxygen in the air.
✓ Understand why carbon monoxide is poisonous, in terms of its effect on the capacity of blood to transport oxygen.
✓ Know that, in car engines, the temperature reached is high enough to allow nitrogen and oxygen from air to react, forming oxides of nitrogen.
✓ Be able to explain how the combustion of some impurities in hydrocarbon fuels results in the formation of sulfur dioxide.
✓ Understand how sulfur dioxide and oxides of nitrogen oxides contribute to acid rain.
✓ Be able to describe how long-chain alkanes are converted to alkenes and shorter-chain alkanes by catalytic cracking (using silica or alumina as the catalyst and a temperature in the range of 600–700 °C).
✓ Be able to explain why cracking is necessary, in terms of the balance between supply and demand for different fractions.

WHAT ARE FOSSIL FUELS?

Crude oil, natural gas and coal are **fossil fuels**.

Crude oil was formed millions of years ago from the remains of animals and plants that were pressed together under layers of rock. It is usually found deep underground, trapped between layers of rock that it can't seep through (impermeable rock). Natural gas is often trapped in pockets above crude oil.

The supply of fossil fuels is limited – they are called finite or **non-renewable** fuels. They are an extremely valuable resource that must be used efficiently.

Crude oil contain many useful **hydrocarbons**, which are molecules that contain only carbon and hydrogen. These hydrocarbons must be separated so that they are not wasted.

Δ Fig. 4.2 The biggest oil refinery in Europe – in the Netherlands.

SEPARATING THE FRACTIONS

The chemicals in crude oil are separated into useful **fractions** by a process known as fractional distillation.

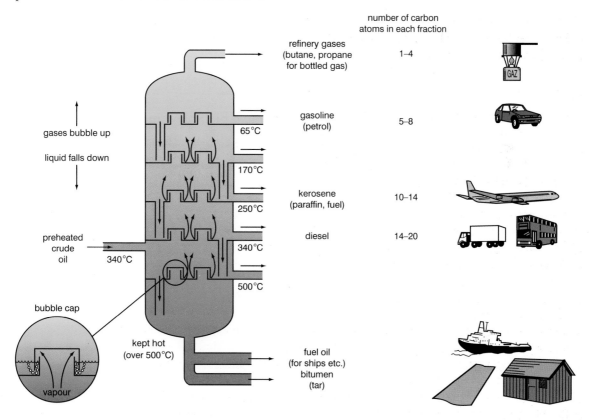

Δ Fig. 4.3 The fractionating column converts the crude oil into many useful fractions.

The crude oil is heated in a furnace and passed into the bottom of a fractionating column. It gives off a mixture of vapours which rise up the column, and the different fractions condense out at different heights. The fractions that come off near the top are light-coloured runny liquids. Those removed near the bottom of the column are dark and sticky. Thick liquids that are not runny, such as these bottom-most fractions, are described as viscous.

HOW DOES FRACTIONAL DISTILLATION WORK?

The components of crude oil separate because they have different boiling points. A simple particle model explains why their boiling points differ. Crude oil is a mixture of hydrocarbon molecules. The molecules are chemically bonded in similar ways with strong covalent bonds (see page 80) but contain different numbers of carbon atoms.

heptane
$$H-\underset{\underset{H}{|}}{\overset{\overset{H}{|}}{C}}-\underset{\underset{H}{|}}{\overset{\overset{H}{|}}{C}}-\underset{\underset{H}{|}}{\overset{\overset{H}{|}}{C}}-\underset{\underset{H}{|}}{\overset{\overset{H}{|}}{C}}-\underset{\underset{H}{|}}{\overset{\overset{H}{|}}{C}}-\underset{\underset{H}{|}}{\overset{\overset{H}{|}}{C}}-\underset{\underset{H}{|}}{\overset{\overset{H}{|}}{C}}-H$$

octane
$$H-\underset{\underset{H}{|}}{\overset{\overset{H}{|}}{C}}-\underset{\underset{H}{|}}{\overset{\overset{H}{|}}{C}}-\underset{\underset{H}{|}}{\overset{\overset{H}{|}}{C}}-\underset{\underset{H}{|}}{\overset{\overset{H}{|}}{C}}-\underset{\underset{H}{|}}{\overset{\overset{H}{|}}{C}}-\underset{\underset{H}{|}}{\overset{\overset{H}{|}}{C}}-\underset{\underset{H}{|}}{\overset{\overset{H}{|}}{C}}-\underset{\underset{H}{|}}{\overset{\overset{H}{|}}{C}}-H$$

△ Fig. 4.4 Octane has one more carbon atom and two more hydrogen atoms than heptane. Their formulae differ by CH_2.

REMEMBER

Remember, the larger the molecule, the stronger the attractive force between the molecules.

The weak attractive forces between the molecules must be broken for the hydrocarbon to boil. The longer a hydrocarbon molecule is, the stronger the intermolecular forces between the molecules. The stronger these forces of attraction, the higher the boiling point, since more energy is needed to overcome the larger forces.

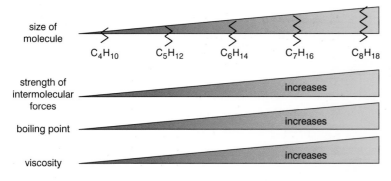

△ Fig. 4.5 How the properties of hydrocarbons change as the molecule gets longer.

The smaller-molecule hydrocarbons are more **volatile**: they more easily form a vapour. For example, we can smell petrol (with molecules containing between 5 and 10 carbon atoms) much more easily than engine oil (with molecules containing between 14 and 20 carbon atoms) because petrol is more volatile.

Another difference between the fractions is how easily they burn and how smoky their flames are.

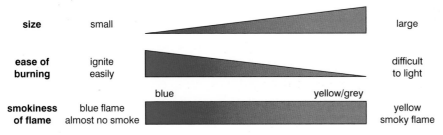

Δ Fig. 4.6 How different hydrocarbons burn.

QUESTIONS

1. Crude oil is a non-renewable fuel. What does this mean?

2. When drilling for oil, there is often excess gas to be burned off. What is this gas? Where does it come from?

3. One of the oil fractions obtained from the fractional distillation of crude oil is light-coloured and runny. Is this fraction more likely to have a small chain of carbon atoms or a long chain?

4. Another of the oil fractions obtained from the fractional distillation of crude oil burns with a very sooty yellow flame. Is this fraction more likely to have a small chain of carbon atoms or a long chain?

5. Some fractions obtained from crude oil are very volatile. What does this mean?

CRACKING THE OIL FRACTIONS

The composition of crude oil varies in different parts of the world. The table shows the composition of a sample of crude oil from the Middle East after fractional distillation.

Fraction (in order of increasing boiling point)	Percentage produced by fractional distillation
liquefied refinery gases	3
gasoline	13
kerosene	12
diesel	14
fuel oil and bitumen	49

Δ Table 4.1 Oil fractions.

Table 4.1 illustrates that the fractional distillation of crude oil produces a large proportion of the higher boiling point fractions and much smaller proportions of the lower boiling point fractions. This is a dilemma as the worldwide demand for the lower boiling point fractions greatly exceeds the demand for the higher boiling point fractions. The process of cracking is used to provide a solution to this issue. The catalytic cracking process converts long-chain alkanes into alkenes (a homologous series of hydrocarbons – see page 269) and shorter-chain alkanes using silica or alumina as the catalyst and a temperature in the range of 600–700 °C.

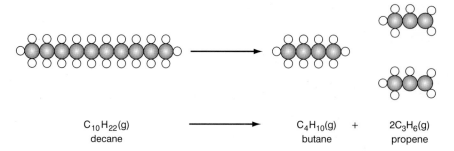

$C_{10}H_{22}(g)$ $C_4H_{10}(g)$ + $2C_3H_6(g)$
decane   butane   propene

Δ Fig. 4.7 The decane molecule ($C_{10}H_{22}$) is converted into the smaller molecules butane (C_4H_{10}) and propene (C_3H_6).

The butane and propene formed in this example of cracking have different types of structures.

REMEMBER

- Propene belongs to a family of hydrocarbons (known as a **homologous series**) called **alkenes** (see page 269).

- Alkenes are much more reactive (and hence useful) than hydrocarbons like decane (an alkane).

Developing investigative skills

A student set up an experiment to 'crack' some liquid paraffin. The student soaked some mineral wool in the liquid paraffin and assembled the apparatus as shown. The student then heated the pottery pieces very strongly, occasionally letting the flame waft onto the mineral wool. Bubbles of gas started to collect in the test tube. After a few minutes the student had collected three test tubes full of gas and so the student stopped heating. Almost immediately, water from the trough started to travel up the delivery tube towards the boiling tube.

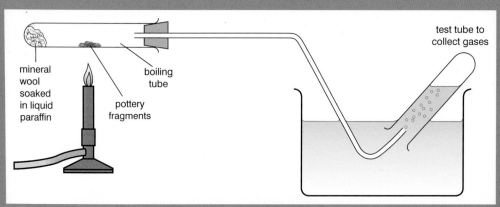

Δ Fig. 4.8 Apparatus for experiment.

Devise and plan investigations

❶ The gas or gases produced in this reaction can be collected by displacement of water. What property of the gas(es) does this demonstrate?

❷ Why did the water start to travel up the delivery tube when heating was stopped?

Demonstrate and describe techniques

❸ What are the hazards involved in this experiment? What safety precautions would minimise them?

❹ The first test tube of gas collected did not burn but the second one did. Explain this difference.

❺ The third test tube of gas decolourised bromine water. What does this suggest about the gas present?

Analyse and interpret data

❻ The student suggested that one of the two products was ethene (C_2H_4). Assuming that liquid paraffin has the formula $C_{14}H_{30}$, write a balanced symbol equation for the cracking of the liquid paraffin.

QUESTIONS

1. The cracking of hydrocarbons often produces ethene. To which homologous series does ethene belong?

2. Explain why cracking is needed in addition to the fractional distillation of crude oil?

3. What conditions are needed for the cracking of oil fractions?

COMBUSTION OF FUELS

Most of the common fuels used today are hydrocarbons. When a hydrocarbon is burned in a plentiful supply of air, it reacts with the oxygen in the air (it is oxidised) to form carbon dioxide and water. This reaction is an example of **combustion**.

hydrocarbon + oxygen → carbon dioxide + water

For example, when methane (natural gas) is burned:

$$CH_4(g) + 2O_2(g) \rightarrow CO_2(g) + 2H_2O(l)$$

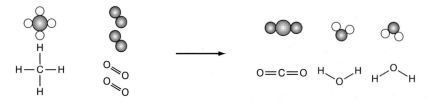

△ Fig. 4.9 The complete combustion of methane in a plentiful supply of air.

The air contains only about 21% oxygen by volume. When a hydrocarbon fuel is burned, there is not always enough oxygen for complete combustion. Instead, some incomplete combustion occurs, forming carbon or carbon monoxide:

methane	+	oxygen	→	carbon monoxide	+	water
$2CH_4(g)$	+	$3O_2(g)$	→	$2CO(g)$	+	$4H_2O(l)$

methane	+	oxygen	→	carbon	+	water
$CH_4(g)$	+	$O_2(g)$	→	$C(s)$	+	$2H_2O(l)$

Incomplete combustion is expensive because the full energy content of the fuel is not being released, and the formation of carbon or soot reduces the efficiency of the burner. It can be dangerous: carbon monoxide is extremely poisonous. It reduces the capacity of the haemoglobin in the blood to carry oxygen. Brain cells deprived of their supply of oxygen die quickly. The sign that a fuel is burning incompletely is that the flame is yellow. When complete combustion occurs, the flame is blue.

Fig. 4.10 The gas in this cooker is burning completely.

Another problem associated with burning petrol or diesel in a car engine is that the temperature is high enough for nitrogen and oxygen to combine to form various oxides of nitrogen. These are sometimes represented by the formula NO_x. These oxides of nitrogen react with water in the atmosphere to form nitric acid, one of the acids responsible for **acid rain** (see also page 252).

QUESTIONS

1. How can you tell if a fuel is burning with an insufficient supply of oxygen?

2. Name two products that can form when methane burns in an insufficient supply of oxygen.

3. How does carbon monoxide act as a poison?

4. Name some alternative ways of generating electricity that do not involve burning fossil fuels.

SULFUR DIOXIDE

Sulfur dioxide is formed when sulfur is burned in oxygen. The combustion of some impurities in hydrocarbon fuels results in the formation of sulfur dioxide.

It reacts with water to form sulfurous acid.

sulfur dioxide	+	water	→	sulfurous acid
$SO_2(g)$	+	H_2O (l)	→	$H_2SO_3(aq)$

In the atmosphere sulfur dioxide, which is a waste product from burning petrol in a car, can react with water and oxygen to form sulfuric acid.

sulfur dioxide	+	oxygen	+	water	→	sulfuric acid
$2SO_2(g)$	+	$O_2(g)$	+	$2H_2O(l)$	→	$2H_2SO_4(aq)$

THE REACTION OF NITROGEN WITH OXYGEN

Nitrogen is usually a very unreactive gas. It does not normally react with oxygen in the air to form oxides. However, under special circumstances it can react with oxygen to form nitrogen monoxide and then nitrogen dioxide. Lightning in thunderstorms can release enough energy for these reactions to take place. Also, car engines can generate a high enough temperature and pressure to make the reactions happen.

nitrogen	+	oxygen	→	nitrogen monoxide
$N_2(g)$	+	$O_2(g)$	→	$2NO(g)$

| nitrogen monoxide | + | oxygen | → | nitrogen dioxide |
| $2NO(g)$ | + | $O_2(g)$ | → | $2NO_2(g)$ |

Nitrogen oxides react with water and oxygen to form nitric acid. This then becomes 'acid rain' that affects forests, aquatic life and buildings.

ACID RAIN

Burning fossil fuels gives off many gases, including sulfur dioxide and various nitrogen oxides.

Sulfur dioxide reacts with water to form sulfurous acid or with water and oxygen to form sulfuric acid. Nitrogen oxide combines with water and oxygen to form nitric acid. These substances can make the rain acidic, hence the name acid rain.

Buildings, particularly those made of limestone and marble (both are forms of calcium carbonate, $CaCO_3$), are damaged by acid rain. Metal structures are also damaged by sulfuric acid.

Acid rain harms plants that take in the acidic water and the animals that live in the affected rivers and lakes. It washes ions such as calcium and magnesium out of the soil, depleting the minerals available to plants. Acid rain also washes aluminium out of the soil and into rivers and lakes, where it poisons freshwater fish.

Reducing emission of the gases that cause acid rain is expensive. Part of the problem is that the acid rain usually falls a long way from the places where the gases were given off.

Power stations are now being fitted with flue gas desulfurisation plants (FGD) to reduce the release of sulfur dioxide into the atmosphere.

Modern cars and trucks have catalytic converters fitted to the exhaust system to reduce the level of pollutants released into the atmosphere from burning petrol.

△ Fig. 4.11 Most lakes in Scandinavia are highly acidic due to acid rain. The acid can be neutralised by spreading slaked lime.

Catalytic converters change emissions such as carbon monoxide (CO) into carbon dioxide and nitrogen oxides (nitrogen(II) oxide, NO) into nitrogen and oxygen. For example:

$2CO(g) + O_2(g) → 2CO_2(g)$

The reduction in nitrogen oxides reduces acid rain and the carbon dioxide produced causes less damaging acid rain. However, the release of more carbon dioxide into the atmosphere may be contributing to global warming.

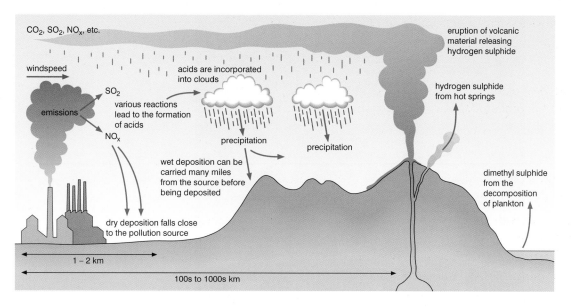

△ Fig. 4.12 How natural and man-made activities contribute to acid rain.

QUESTIONS

1. How does sulfur dioxide form sulfuric acid in the atmosphere?

2. How does nitrogen oxide form nitric acid in the atmosphere?

3. What can be added to the water in lakes to reduce levels of acidity?

SCIENCE IN CONTEXT

THE FOSSIL FUEL DILEMMA

There is widespread agreement that the supplies of the non-renewable fossil fuels – oil, gas and coal – will eventually run out. However, it is not easy to estimate exactly when these supplies will run out. Many different factors need to be considered, including how much of each deposit is left in the Earth, how fast we are using each fossil fuel at the moment, whether countries that have supplies will sell to those that don't, and how this is likely to change in the future. If we start switching to alternative fuel sources that are renewable, the reserves that we have will last longer.

Current estimates suggest that crude oil will run out between 2025 and 2070. The estimate for natural gas is similar, with 2060 a possible date. The situation with coal is very different. Most coal deposits have not yet been tapped, and the decline of the coal mining industry in countries such as the UK means that many coal seams are currently lying undisturbed. If we carry on using coal at the same rate as we do

△ Fig. 4.13 A coal-fired power station.

today, we could have enough coal to last well over a thousand years. However, as other fossil fuels run out, particularly oil, the use of coal may increase, reducing that timespan considerably.

So should we increase our efforts to develop renewable forms of energy such as wind and solar energy; should we put greater emphasis on nuclear power; or should we plan to make much greater use of coal? Perhaps we should do all three? Solving this dilemma is likely to depend as much on political decisions as scientific ones. What would you recommend?

End of topic checklist

A **hydrocarbon** is a compound containing hydrogen and carbon only.

Fractional distillation is a process for separating liquids with different boiling points.

Catalytic cracking is the process by which long-chain hydrocarbons (alkanes) are broken down to form more useful short-chain alkanes and alkenes using high temperatures and a catalyst.

The facts and ideas that you should know and understand by studying this topic:

○ Understand that crude oil is a mixture of hydrocarbons.

○ Be able to describe and explain how the industrial process of fractional distillation separates crude oil into fractions.

○ Know the names and uses of the main fractions obtained from crude oil:

- refinery gases
- gasoline
- kerosene
- diesel
- fuel oil
- bitumen.

○ Be able to describe the trend in colour, boiling point and viscosity of the main fractions.

○ Know that combustion of a fuel releases heat energy, and know the products of complete combustion of hydrocarbons with oxygen in the air.

○ Know the possible products of incomplete combustion of hydrocarbons with oxygen in the air.

○ Be able to explain that carbon monoxide is poisonous because it reduces the capacity of the blood to carry oxygen.

○ Know that in car engines the temperature is hot enough for nitrogen and oxygen to combine and form nitrogen oxides.

○ Be able to explain how the combustion of some impurities in hydrocarbon fuels results in the formation of sulfur dioxide.

○ Understand how sulfur dioxide and oxides of nitrogen contribute to acid rain.

○ Be able to explain that the fractional distillation of crude oil produces more long-chain hydrocarbons than can be used and fewer short-chain hydrocarbons than are needed and so cracking is necessary.

○ Be able to describe how long-chain hydrocarbons can be converted into shorter chain hydrocarbons by catalytic cracking.

○ Know that oil fractions are cracked using catalysts of silica or alumina at a temperature of between 600 and 700 °C.

End of topic questions

1. a) How is crude oil formed? **(2 marks)**

b) Why is crude oil a non-renewable fuel? **(1 mark)**

2. The diagram shows a column used to separate the components present in crude oil.

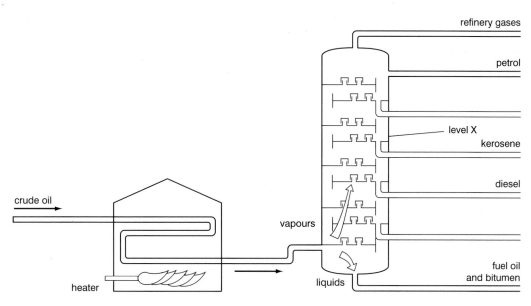

a) Name the process used to separate crude oil into fractions. **(1 mark)**

b) What happens to the boiling point of the mixture as it goes up the column? **(1 mark)**

c) The mixture of vapours arrives at level X. What now happens to the various parts of the mixture? **(2 marks)**

3. The cracking of decane molecules is shown by the equation $C_{10}H_{22} \rightarrow Y + C_2H_4$.

a) Decane is a hydrocarbon. What is a hydrocarbon? **(1 mark)**

b) What reaction conditions are needed for cracking? **(2 marks)**

c) Write down the molecular formula for hydrocarbon Y. **(1 mark)**

d) What homologous series does hydrocarbon Y belong to? **(1 mark)**

e) Why is the cracking of crude oil fractions so important? **(2 marks)**

4. Petrol is a hydrocarbon with a formula of C_8H_{18}.

    a) What are the products formed when petrol burns in a plentiful supply of air?

                  (2 marks)

    b) Write the balanced symbol equation, including state symbols, for the reaction when petrol burns in a plentiful supply of air.   **(2 marks)**

    c) When petrol is burned in a car engine, carbon monoxide may be formed. Explain why carbon monoxide is dangerous.   **(2 marks)**

5. This question is about acid rain.

    a) Name two pollutant gases that contribute to acid rain.   **(2 marks)**

    b) Describe two damaging effects of acid rain.   **(2 marks)**

    c) Explain two ways of reducing the impact of acid rain.   **(2 marks)**

Alkanes

INTRODUCTION

The simplest family or 'homologous series' of **organic molecules** is the alkanes. The first alkane, methane, is the major component of natural gas, a common fossil fuel. Other alkanes are obtained from crude oil and are widely used as fuels.

△ Fig. 4.14 Excess gas is burned as natural gas or oil is extracted from beneath the ocean floor.

ALKANES

Alkanes are made up of carbon atoms linked together by only single covalent bonds and are known as **saturated** hydrocarbons. The **molecular** and **displayed formulae** for alkanes are shown in Table 4.2. The displayed formula shows the bonds, the full arrangement of the atoms and all the bonds between the various atoms. The structural formula gives more detail than the molecular formula but less than the displayed formula – it shows the formula of the sections that make up the molecule and will always show the functional group.

Many alkanes are obtained from crude oil by fractional distillation. The smallest alkanes are used extensively as fuels. Apart from **combustion**, however, they are remarkably unreactive.

Alkane	Molecular formula	Displayed formula	Structural formula	Boiling point (°C)	State at room temperature and pressure
methane	CH_4	H–C–H (with H above and below)	CH_4	−162	gas
ethane	C_2H_6	H–C–C–H	CH_3CH_3	−89	gas
propane	C_3H_8	H–C–C–C–H	$CH_3CH_2H_3$	−42	gas
butane	C_4H_{10}	H–C–C–C–C–H	CH_3CH_2 CH_2CH_3	0	gas
pentane	C_5H_{12}	H–C–C–C–C–C–H	CH_3CH_2 CH_2CH_2 CH_3	36	liquid

△ Table 4.2 Alkanes.

Note: the boiling points of the alkanes increase as the relative molecular mass increases, as is generally the case for all simple molecular structures.

HOMOLOGOUS SERIES

Alkanes form a **homologous series**. Members of a homologous series have certain things in common:

- They have the same **general formula**. For alkanes this is C_nH_{2n+2}
- They have similar chemical properties.
- They show a gradual change in physical properties, such as melting point and boiling point.
- They differ from the previous member of the series by $-CH_2-$.

△ Fig. 4.15 Formula 1 cars use specially blended mixtures of alkane hydrocarbons.

- They contain the same **functional group** (the part of the molecule that is responsible for the similar chemical properties) – in this case, although alkanes only contain C–C and C–H bonds these are not usually considered to be functional groups as they are present in all organic compounds.

PROPERTIES OF ALKANES

The properties of alkanes are given in Table 4.3.

	Alkanes
General formula	C_nH_{2n+2}
Description	saturated (no double C=C bond)
Combustion	burn in oxygen to form CO_2 and H_2O (CO if low supply of oxygen)
Reactivity	low
Chemical test	none
Uses	fuels

△ Table 4.3 Properties of alkanes.

△ Fig. 4.16 Methane is burning in the oxygen in the air to form carbon dioxide and water.

1. Alkanes are saturated hydrocarbons.

   a) What is meant by the word 'saturated'?

   b) What is meant by the word 'hydrocarbon'?

2. **a)** What is the molecular formula for the alkane with five carbon atoms?

   b) What products would you expect to be formed if this alkane were burned in a plentiful supply of oxygen?

3. Would you expect hexane to be a solid, liquid or gas? Explain your answer.

4. What is the functional group for the alkanes?

THE SHAPE OF ALKANE MOLECULES

In alkanes, the four bonds on each carbon atom are directed to the corners of a tetrahedron.

The methane molecule is tetrahedral (four-sided) in shape.

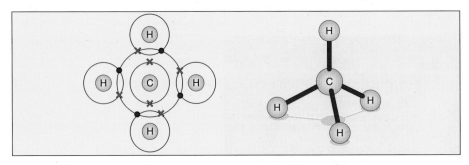

Δ Fig. 4.17 The bonding and structure of methane.

NAMING ALKANES

There is a worldwide agreement on how organic compounds are named. The system used is the one developed by the International Union of Pure and Applied Chemistry (IUPAC). The names of hydrocarbons are based on the number of carbon atoms in their molecules. The first part of the name tells you the number of carbon atoms, for example 'meth-' (from methane) has one carbon atom, 'eth-' (from ethane) has two carbon atoms, 'prop-' (from

Δ Fig. 4.18 The bonding and structure of octane.

propane) has three carbon atoms, and so on. This is the structure of butane (C_4H_{10}):

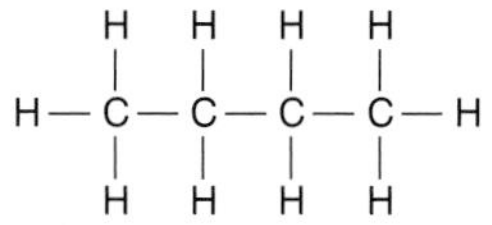

△ Fig. 4.19 Butane

ISOMERS

The carbon atoms in a hydrocarbon molecule can be arranged in different ways. For example, in the alkane with the molecular formula C_4H_{10}, the carbon atoms can be positioned in two ways, while keeping the same molecular formula:

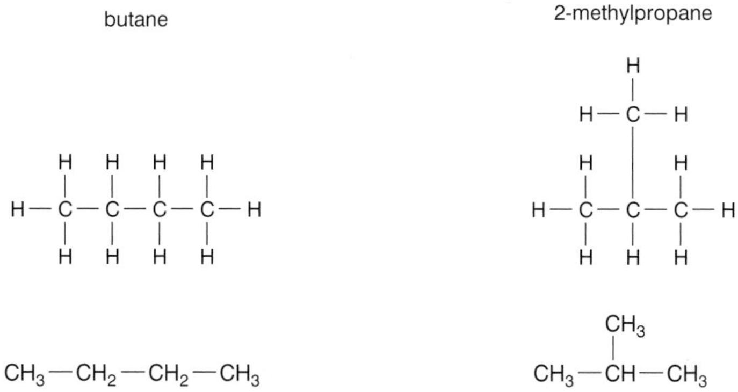

△ Fig. 4.20 Isomers of C_4H_{10}.

An **isomer** is a compound with the same molecular formula, but different structures, to one or more other compounds.

2-methylpropane is a **structural isomer** (same atoms but rearranged) of butane, which has a longer chain of carbon atoms. This feature of alkane structure is called structural isomerism.

The table shows the three isomers of the alkane C_5H_{12}.

Isomer	pentane	2-methylbutane	2,2-dimethylpropane
Structure (displayed formula)	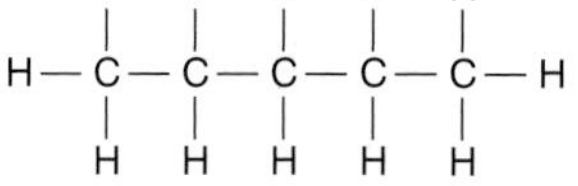	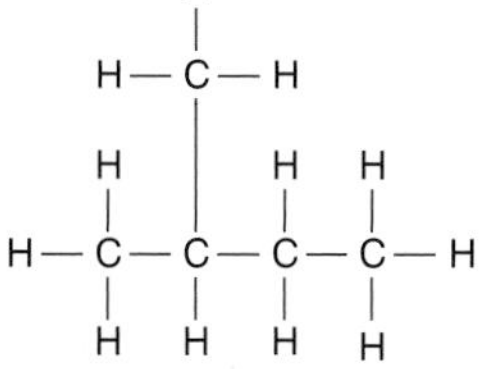	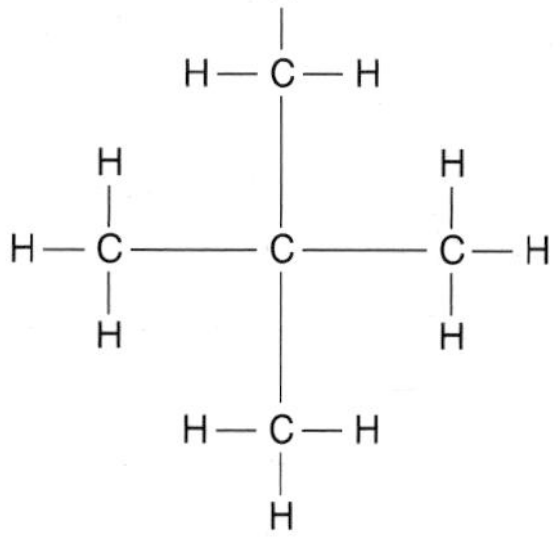
Boiling point (°C)	36	27	11

△ Table 4.4 Isomers of pentane.

1. What are isomers?

2. Write down the structural and displayed formulae of hexane, the alkane with six carbon atoms. What is its molecular formula?

3. Draw the structural and displayed formulae of two different isomers of C_6H_{14}.

COMBUSTION OF ALKANES

In a plentiful supply of air, alkanes will burn to form carbon dioxide and water. A blue flame, as can be produced by a Bunsen burner, indicates complete combustion:

methane	+	oxygen	$\rightarrow$	carbon dioxide	+	water
$CH_4(g)$	+	$2O_2(g)$	$\rightarrow$	$CO_2(g)$	+	$2H_2O(l)$

When the oxygen supply is limited, as when a Bunsen burner burns with a yellow flame, incomplete combustion occurs:

methane	+	oxygen	$\rightarrow$	carbon	+	water
$CH_4(g)$	+	$O_2(g)$	$\rightarrow$	$C(s)$	+	$2H_2O(l)$

The incomplete combustion of hydrocarbons such as methane can be very dangerous. Carbon monoxide, which is extremely poisonous, can be produced. Gas and oil heaters or boilers need to be serviced regularly. This is to ensure that jets do not become blocked and limit the air supply, or exhaust flues become blocked and allow small quantities of carbon monoxide to enter the room. The flame in such a boiler or heater should always be blue in colour.

methane	+	oxygen	$\rightarrow$	carbon monoxide	+	water
$2CH_4(g)$	+	$3O_2(g)$	$\rightarrow$	$2CO(g)$	+	$4H_2O(l)$

QUESTIONS

These questions are based on combustion reactions.

1. Alkanes are useful as fuels. What is the reason for this?

2. Explain why incomplete combustion can be dangerous if the products of this type of combustion are inhaled.

3. Describe how the products of combustion of a short chain hydrocarbon differ from a longer chain hydrocarbon.

THE REACTION OF METHANE WITH BROMINE

The alkanes undergo **substitution reactions** with the halogens, in which a hydrogen atom in the alkane is replaced by a halogen atom. For example, when methane reacts with bromine in the presence of UV light, a substitution reaction occurs and bromomethane is formed.

methane	+ bromine	$\rightarrow$	bromomethane	+	hydrogen bromide
$CH_4(g)$	+ $Br_2(g)$	$\rightarrow$	$CH_3Br(g)$	+	$HBr(g)$

If there is too much bromine, substitution will continue, forming dibromomethane (CH_2Br_2), tribromomethane ($CHBr_3$) and tetrabromomethane (CBr_4).

SCIENCE IN CONTEXT: SOME INTERESTING FACTS ABOUT METHANE

△ Fig. 4.21 A commercial biodigester.

1. Methane makes up about 97% of natural gas.

2. It is formed by the decay of plant matter where there is no oxygen (anaerobic decay).

3. Biogas contains 40–70% methane. Biodigesters convert organic wastes into a nutrient-rich liquid fertiliser and biogas, a renewable source of electrical and heat energy. They are widely used in developing countries, particularly India, Nepal and Vietnam. Biodigesters can help families by providing a cheap source of fuel, preventing environmental pollution from the runoff from animal pens, and reducing diseases caused by the use of untreated manure as fertiliser. However, biodigesters only work efficiently in hot countries; they are not effective at low temperatures.

4. Methane is one of the greenhouse gases, thought by some scientists to be responsible for global warming. It has almost 25 times the effect of the same volume of carbon dioxide.

5. Ruminant animals such as cows and sheep produce methane. It has been estimated that a cow can produce as much as 200 litres of methane per day. So could cows be one of the causes of global warming?

End of topic checklist

A **homologous series** is a group of organic compounds with the same general formula, similar chemical properties and physical properties that change gradually from one member to the next.

A **structural formula** shows the main clusters of atoms in the molecule and the functional group.

A **displayed formula** shows the position of all bonds and atoms in the molecule.

A **saturated** hydrocarbon contains only C–C single bonds.

Isomers are compounds with the same molecular formula but different structural formulae.

A **substitution reaction** is one in which one atom is replaced by an atom of another element.

The facts and ideas that you should know and understand by studying this topic:

○ Know that alkanes have a general formula of C_nH_{2n+2}.

○ Be able to name the first six straight chain members of the alkane homologous series (methane, ethane, propane, butane, pentane, hexane) and understand that these and other hydrocarbons are named using a standard system, the International Union of Pure and Applied Chemistry (IUPAC) rules.

○ Be able to draw structural and displayed formulae for the first five members of the alkane homologous series.

○ Know that alkanes undergo substitution reactions with halogens in the presence of UV light, for example the reaction of methane with bromine to form bromomethane (CH_3Br).

End of topic questions

1. What is meant by each of the following?

   a) Homologous series (1 mark)

   b) Structural isomerism (1 mark)

2. What is the molecular formula for an alkane with ten carbon atoms? (1 mark)

3. Is the compound with the chemical formula $C_{15}H_{30}$ a member of the alkane series? Explain your answer. (1 mark)

4. a) Draw a displayed formula for pentane, an alkane with five carbon atoms. (1 mark)

   b) Draw a displayed formula for another isomer of pentane. (1 mark)

   c) Explain why octane (C_8H_{18}) is a liquid whereas pentane is a gas. (1 mark)

   d) Is octane a saturated or unsaturated hydrocarbon? Explain your answer. (1 mark)

5. Ethane burns in excess oxygen.

   a) What is the molecular formula of ethane? (1 mark)

   b) Name the products formed when ethane burns in excess oxygen. (2 marks)

   c) What colour flame would indicate that the ethane was burning in excess oxygen? (1 mark)

   d) Write a balanced symbol equation for the reaction. (2 marks)

   e) Name two additional products that could be formed if the oxygen supply was limited. (2 marks)

   f) What colour flame would indicate that the ethane was burning in a limited supply of oxygen? (1 mark)

6. Methane will react with the halogen chlorine.

   a) What type of reaction is this? (1 mark)

   b) What reaction conditions are needed? (1 mark)

Alkenes

INTRODUCTION

Alkenes are hydrocarbons and burn in air in the same way that alkanes do. However, in comparison to alkanes, alkenes are much more reactive due to the carbon=carbon double bond they contain. This makes them very useful starting materials for a number of important industrial processes, including the manufacture of synthetic polymers and margarine.

△ Fig. 4.22 Margarine is made from vegetable oil whose unsaturated molecules have been saturated with hydrogen.

KNOWLEDGE CHECK

✓ Understand the sharing of electrons in a double covalent bond.
✓ Know the typical physical properties of compounds that exist as simple molecules.
✓ Know the simple properties of alkanes.

LEARNING OBJECTIVES

✓ Know that alkenes contain the functional group C=C and general formula C_nH_{2n}.
✓ Be able to explain why alkenes are classified as unsaturated hydrocarbons.
✓ Be able to name alkenes with up to six carbon atoms using the rules of the International Union of Pure and Applied Chemistry (IUPAC).
✓ Understand how to draw the structural and displayed formulae for alkenes with up to four carbon atoms in the molecule, and name the unbranched-chain isomers.
✓ Be able to describe the reactions of alkenes with bromine to produce dibromoalkanes.
✓ Be able to describe how bromine water can be used to distinguish between an alkane and an alkene.

PROPERTIES OF ALKENES

Alkenes are another homologous series, so they have similar chemical properties and physical properties that change gradually from one member to the next. The functional group for alkenes is a double C=C bond.

Alkene	Molecular formula	Displayed formula		Structural formula	Boiling point (°C)	State at room temperature and pressure
ethene	C_2H_4	ethene	H₂C=CH₂ structure	$CH_2=CH_2$	−104	gas
propene	C_3H_6	propene	H—C—C=C structure	$CH_3CH=CH_3$	−48	gas
butene	C_4H_8	butene	H—C—C—C=C structure	CH_3CH_2 $CH=CH_2$	−6	gas

△ Table 4.5 Alkenes.

Notice again that the boiling point increases as the relative molecular mass increases. Alkenes are often formed by the catalytic **cracking** of larger hydrocarbons (see page 247). They contain one or more carbon=carbon double bonds. Hydrocarbons with at least one double bond are known as **unsaturated** hydrocarbons. This is in contrast to the alkanes which are saturated as they only contain C–C single bonds. Alkenes burn well and are reactive in other ways also. Their reactivity is due to the carbon=carbon double bond.

A simple test to tell alkenes from alkanes is by adding bromine water to the hydrocarbon. Alkanes do not react with bromine water, so the colour does not change. An alkene reacts with the bromine so the bromine water loses its colour. The type of reaction is known as an **addition reaction**:

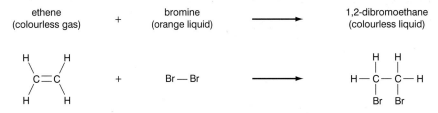

△ Fig. 4.23 The reaction of ethene and bromine.

The bromine molecule (Br_2) splits and the two bromine atoms add on to the carbon atoms on either side of the double bond. The 1,2–dibromoethane is an example of a dibromoalkane. Dibromoalkanes are always the product of an alkene reacting with bromine.

Properties of alkenes	
General formula	C_nH_{2n}
Description	Unsaturated (contains a double C=C bond)
Combustion	Burn in oxygen to form CO_2 and H_2O (CO if low supply of oxygen)
Reactivity	High (because of double C=C bond); undergo addition reactions

Properties of alkenes	
Chemical test	Turn bromine water from orange to colourless (an addition reaction)
Uses	Making polymers (addition reactions) such as poly(ethene)

△ Table 4.6 Properties of alkenes.

Alkenes can also form isomers.

For example, the isomers of butene (C_4H_8) are:

but-1-ene   but-2-ene

△ Fig. 4.24 But-1-ene and but-2-ene.

QUESTIONS

1. Ethene is an unsaturated hydrocarbon. What does 'unsaturated' mean?

2. Name a large-scale use of ethene.

3. This question is about C_3H_6.

   a) Draw a structural formula for the C_3H_6 molecule.

   b) Draw a displayed formula for the C_3H_6 molecule.

SCIENCE IN CONTEXT

SATURATED AND UNSATURATED FATS

We all need some fat in our diet because it helps the body to absorb certain nutrients. Fat is also a source of energy and provides essential fatty acids. However, it is best to keep the amount of fat we eat at sensible levels and to eat **unsaturated** fats (containing C=C bonds) rather than **saturated** fats (containing C–C bonds) whenever possible. A diet high in saturated fat can cause the level of cholesterol in the blood to build up over time. Raised cholesterol levels increase the risk of heart disease.

Foods high in saturated fat include:

- fatty cuts of meat
- meat products and pies

△ Fig. 4.25 This oil contains unsaturated fats.

- butter
- cheese, especially hard cheese
- cream and ice cream
- biscuits and cakes.

Unsaturated fat is found in:

- oily fish such as salmon, tuna and mackerel
- avocados
- nuts and seeds
- sunflower and olive oils.

So having a carbon=carbon double bond does make a difference!

EXTENSION

Most of all the compounds discovered in the world today are organic. There are so many due to the ability of carbon to form different types of bonds. Alkanes and alkenes are just two examples of homologous series of compounds from the vast number of organic compounds known to date. Use your knowledge of basic organic chemistry to answer the questions below.

1. Alkanes have a general formula C_nH_{2n+2} and alkenes C_nH_{2n}. When $n = 5$ the alkane formed is pentane and the alkene pentene.

   a) Alkanes and alkenes are both hydrocarbons. State why alkanes are considered to be saturated hydrocarbons whereas alkenes are unsaturated.

   b) Give the structural formulae for the alkane and alkene where $n = 6$ (hexane and hexene). Draw a displayed formula for hexane and hexene and label the different types of bonds.

   c) Hexane can have a number of isomers. Draw an isomer of hexane and suggest a name for this isomer.

   d) Although both these compounds have six carbon atoms they undergo different types of reactions with bromine. Name the type of reaction that hexane and hexene will undergo with bromine and draw one product for each reaction.

End of topic checklist

An **unsaturated** hydrocarbon contains C=C double bonds.

An **addition reaction** involves the reaction of an alkene and another element or compound to form a single compound.

The facts and ideas that you should know and understand by studying this topic:

◯ Know that alkenes have a general formula of C_nH_{2n}.

◯ Be able to name the first five straight chain members of the alkene homologous series (ethene, propene, butene, propene, hexene).

◯ Be able to draw structural and displayed formulae for the first three members of the alkene homologous series, including the isomers of butene.

◯ Be able to describe the reactions of alkenes with bromine to produce dibromoalkanes, and classify these as addition reactions.

◯ Be able to describe the reaction of an alkene with bromine water and know that this reaction can be used to identify alkenes.

End of topic questions

1. A hydrocarbon has the formula C_4H_8.

   a) Is the hydrocarbon saturated or unsaturated? Explain your answer. **(1 mark)**

   b) What is the name of the hydrocarbon? **(1 mark)**

   c) Draw displayed formulae of two different isomers of this hydrocarbon.

   (2 marks)

2. A molecule has the molecular formula C_3H_6.

   a) What is its general formula? **(1 mark)**

   b) What is its functional group? **(1 mark)**

3. Ethene burns in oxygen.

   a) Name the products formed when there is a plentiful supply of oxygen.

   (2 marks)

   b) i) Write a balanced symbol equation for the burning of ethene in a plentiful supply of oxygen. **(2 marks)**

   ii) What colour would the flame be? **(1 mark)**

   c) When ethene is burnt in a limited supply of air carbon and water are formed.

   i) Write a balanced symbol equation for this reaction. **(2 marks)**

   ii) What colour would the flame be? **(1 mark)**

4. a) Write molecular formulae for butane and butene. **(2 marks)**

   b) Which hydrocarbon is unsaturated? **(1 mark)**

   c) Which substance could you use to distinguish between butane and butene?

   (1 mark)

5. Propene gas is bubbled through some bromine water.

   a) Describe the colour change that would occur in the bromine water. **(2 marks)**

   b) Write a balanced symbol equation for this reaction. **(2 marks)**

   c) What type of chemical reaction is this an example of? **(1 mark)**

6. In a healthy diet which type of fat is more healthy, a saturated fat or an unsaturated fat? **(1 mark)**

Alcohols

INTRODUCTION

Alcohols are another homologous series of organic compounds with ethanol as the most common alcohol. There are many more uses for ethanol than for other alcohols, and so its manufacture is very important. Ethanol may be produced by a 'natural' method starting with sugar, and by a 'synthetic' method starting with a product of crude oil.

△ Fig. 4.26 The ethanol being made here could be used as a disinfectant in hospitals or as a fuel in cars.

KNOWLEDGE CHECK

✓ Understand the term 'homologous series'.
✓ Know the typical physical properties of compounds that exist as simple molecules.
✓ Know what isomers are.

LEARNING OBJECTIVES

✓ Know that alcohols contain the functional group –OH.
✓ Be able to name alcohols with up to six carbon atoms, using the rules of the International Union of Pure and Applied Chemistry (IUPAC).
✓ Be able to to draw structural and displayed formulae for methanol, ethanol, propanol (propan-1-ol only) and butanol (butan-1-ol only), and name each compound.
✓ Know that ethanol can be oxidised by:
  • burning in air or oxygen (complete combustion)
  • reaction with oxygen in the air to form ethanoic acid (microbial oxidation)
  • heating with potassium dichromate(VI) in dilute sulfuric acid to form ethanoic acid.
✓ Know that ethanol can be manufactured by:
  • reacting ethene with steam in the presence of a phosphoric acid catalyst at a temperature of about 300 °C and a pressure of about 60–70 atm
  • the fermentation of glucose, in the absence of air, at an optimum temperature of about 30 °C and using the enzymes in yeast.
✓ Understand the reasons for fermentation, in the absence of air, and at an optimum temperature.

ALCOHOLS

Alcohols are molecules that contain the –OH **functional group**, which is responsible for their properties and reactions.

Alcohols have the general formula $C_nH_{2n+1}OH$ and belong to the same homologous series, part of which is shown below. It is important to be able to name alcohols with up to six carbon atoms. For example, $C_6H_{13}OH$ is hexanol.

Alcohol	Formula	Displayed formula	Boiling point (°C)
Methanol	CH_3OH	H—C—OH with H above, H below	65
Ethanol	C_2H_5OH	H—C—C—OH with H atoms	78
Propanol	C_3H_7OH	H—C—C—C—OH with H atoms	97
Butanol	C_4H_9OH	H—C—C—C—C—OH with H atoms	118

△ Table 4.7 Properties of alcohols.

ETHANOL – THE MOST COMMON ALCOHOL

Ethanol, commonly just called 'alcohol', is the most widely used of the alcohol family. Its major uses are given in Table 4.8.

Use of ethanol	Reason
Solvent, such as disinfectants and perfumes	The –OH group allows it to dissolve in water, and it dissolves other organic compounds.
Fuel, such as for cars	It releases fewer pollutant gases than petrol (less NO_x and SO_2). It is a renewable resource because it comes from plants, for example sugar beet and sugar cane.
Alcoholic drinks, such as wine, beer, spirits	Alcohol is a depressant and slows down reactions. Some people like this effect. However, alcohol can potentially damage the liver and other organs in the body.

△ Table 4.8 Uses of ethanol.

▷ Fig. 4.28 Brazilians use *alcool* as vehicle fuel.
It is made from the fermented and distilled
juice of sugar cane.

QUESTIONS

1. What would be the formula of the alcohol butanol, which has
four carbon atoms?

2. What are the advantages of using ethanol as a fuel in a
motor car?

3. Ethanol is commonly used as a solvent. What is a solvent?

MANUFACTURING ETHANOL BY FERMENTATION

Ethanol is made by **fermentation**. This involves mixing a sugar
solution with yeast and maintaining the temperature between 25 and
30 °C in the absence of air. The yeast contains **enzymes**, which
catalyse (speed up) the breaking down of the sugar. Enzymes are not
effective if the temperature is too low, and they are destroyed
(denatured) if the temperature is too high.

Fermentation of sugar takes place in large vats. However, even with the
yeast as a catalyst, the process is slow and it takes several days to be
completed. As the concentration of the ethanol increases, the activity
of the yeast decreases, so eventually fermentation stops.

The chemical reaction for fermentation is:

$$\text{sugar} \xrightarrow{\text{yeast}} \text{ethanol} + \text{carbon dioxide}$$

$$C_6H_{12}O_6\,(aq) \xrightarrow{\text{yeast}} 2C_2H_5OH\,(l) + 2CO_2\,(g)$$

Fermentation takes time because it is an enzymic reaction (yeast) and
a batch process. At the end, a more concentrated solution of ethanol is
extracted by fractional distillation. The mixture is boiled and the
ethanol vapour reaches the top of the fractionating column, where it
condenses back to a liquid.

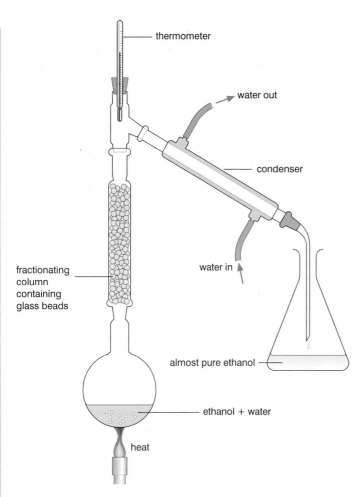

Δ Fig. 4.29 Laboratory apparatus for fractional distillation of an ethanol and water mixture. In large-scale manufacture different equipment would be used but the principles of separation are the same.

QUESTIONS

1. What does the word 'fermentation' mean?

2. What is the optimum temperature for fermentation?

3. Why is yeast needed in the fermentation process?

4. What other compound (as well as ethanol) is produced in the fermentation process?

Developing investigative skills

A student needed to carry out the fermentation of sugar in the laboratory to make some ethanol for fuel. The student set up the apparatus as shown in the diagram below. At first there seemed to be nothing happening, but by the next morning the reaction had started. When the reaction had finished, the student used fractional distillation to obtain some ethanol. (Note: ethanol is highly flammable.)

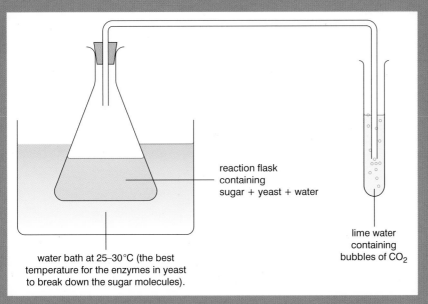

reaction flask
containing
sugar + yeast + water

lime water
containing
bubbles of CO_2

water bath at 25–30°C (the best temperature for the enzymes in yeast to break down the sugar molecules).

△ Fig. 4.30 Fermentation apparatus.

Devise and plan investigations

❶ How could the student have tried to maintain the temperature of the water bath while the reaction was proceeding?

❷ What was the purpose of the lime water?

❸ What would the student have observed to indicate that a reaction had started?

❹ How would the student know when the reaction was complete?

❺ In the fractional distillation process, which liquid would be collected first, ethanol or water? Explain your answer.

MANUFACTURING ETHANOL FROM ETHENE

On an industrial scale, ethanol is also made from ethene, which is obtained from crude oil. The reaction is:

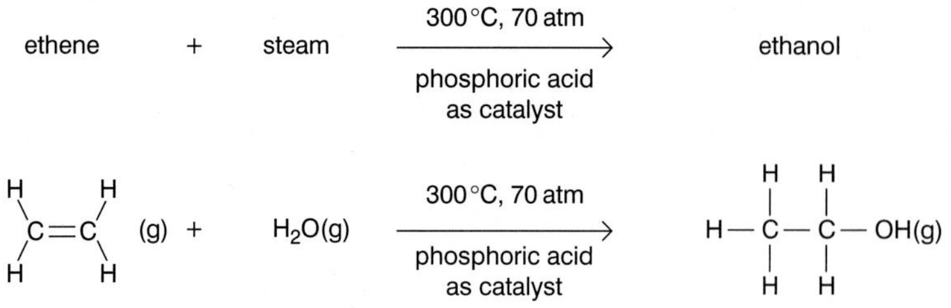

ethene   +   steam $\xrightarrow[\text{phosphoric acid as catalyst}]{\text{300°C, 70 atm}}$ ethanol

$$\underset{H}{\overset{H}{>}}C=C\underset{H}{\overset{H}{<}} \;(g) + H_2O(g) \xrightarrow[\text{phosphoric acid as catalyst}]{\text{300°C, 70 atm}} H-\underset{H}{\overset{H}{\underset{|}{\overset{|}{C}}}}-\underset{H}{\overset{H}{\underset{|}{\overset{|}{C}}}}-OH(g)$$

△ Fig. 4.31 Reaction of ethene and steam to make ethanol.

These are quite extreme conditions in terms of energy (300 °C) and specialist plant equipment (to generate 70 atmospheres), so the process is expensive.

Unlike fermentation, this process is continuous and produces ethanol at a fast rate.

THE CHOICE OF METHOD IN THE MANUFACTURE OF ETHANOL

The best method for making ethanol depends on local circumstances: for example, how much sugar cane or crude oil is available.

	Advantage	Disadvantage
Fermentation	Uses renewable resources. Produces flavour of alcoholic drinks.	Slow 'batch' process. Only small amount of ethanol produced.
Ethene + steam	Fast continuous processes. Large amounts of ethanol produced.	Uses non-renewable resource.

△ Table 4.9 Advantages and disadvantages of methods of making ethanol.

Ethanol undergoes oxidation reactions and, depending on the reaction conditions can produce different products.

1. Combustion in air or oxygen

When the combustion is complete (with sufficient oxygen) the products are carbon dioxide and water.

$$C_2H_5OH(l) + 3O_2(g) \rightarrow 2CO_2(g) + 3H_2O(l)$$

2. Microbial oxidation

In the fermentation process to make ethanol from sugar, if the yeast is contaminated by other microbes the ethanol produced can be oxidised into ethanoic acid (CH_3COOH, acetic acid, vinegar). Most microbial action takes place when wine is exposed to air after fermentation. This can happen in the wine making process and makes the wine undrinkable.

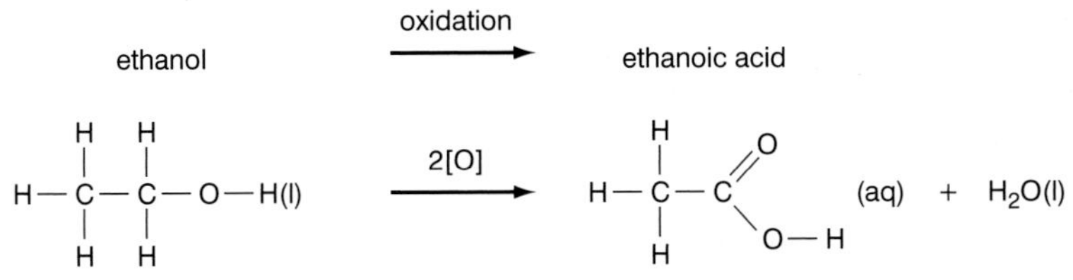

△ Fig. 4.32 Oxidising ethanol to make ethanoic acid.

3. Oxidation by potassium dichromate(VI)

If ethanol is heated with potassium dichromate in the presence of dilute sulfuric acid the ethanol is oxidised to ethanoic acid. This is the reaction that is commonly used to prepare ethanoic acid in the laboratory.

Ethanoic acid belongs to an important homologous series of carboxylic acids (see the next topic). The carboxylic acids are important starting materials for making esters and polyesters (see later topics).

SCIENCE IN CONTEXT

ETHANOL AS A FUEL

The development of ethanol as a fuel for cars has a long history. In the USA cars made by Ford (especially the Model T) used ethanol as a fuel until 1908. Now most cars in the USA can run on blends of petrol/ethanol containing up to 10% ethanol. Petrol in the European Union has up to 10% ethanol.

△ Fig. 4.33 In the USA, the ethanol in gasohol is made from maize.

Brazil and the USA are the world's top producers of ethanol. In Brazil about one-fifth of all cars can use 100% ethanol fuel, known as E100; many other cars use petrol/ethanol blends. Brazil makes its ethanol mostly from fermented sugar obtained from sugar cane; in the USA corn is used as the source of the sugar. Starting from sugar cane is much more efficient, and so ethanol is much cheaper to produce in Brazil than in the USA.

One problem with using ethanol as a fuel is that it readily absorbs water from the atmosphere. This causes some problems when transporting the fuel and makes it more expensive than petrol to transport.

A key question is, should food be used to make a fuel? Is it better to use fuel made from crude oil? Some people argue that the food is needed for people to eat and should not be used to run cars. They say that using so much land for growing sugar cane and corn to make ethanol means that less land is available for growing food. This leads to food shortages and increases in the price of food. What do you think?

Ethanol is an alcohol with the formula C_2H_5OH. It is used for a wide range of purposes, from producing medicine, synthesising chemical products, as a fuel as well as a beverage. Fermentation of sugar into ethanol was one of the earliest organic reactions carried out and distillation was well known by the early Greeks. Ethanol is now manufactured from the products of oil. For the questions below use your knowledge of organic chemistry and the work you have studied on rates and equilibria to understand the different factors affecting the manufacture of ethanol in today's society.

1. Ethanol can be produced by the direct hydration of ethene. The conditions for the manufacture are 300 °C, 60–70 atmospheres and phosphoric(V) acid as a catalyst.

   a) The reaction of ethene and steam in the formation of ethanol is reversible and the reaction is exothermic. Write a balanced symbol equation to show this and describe what is meant by exothermic.

   b) Consider the temperature. In order to produce the maximum possible amount of ethanol in the equilibrium mixture an 'optimum temperature' or 'compromise temperature' is needed. Explain why the manufacturers do not use a very high temperature.

   c) Consider the pressure. Use your knowledge of the effects of changing the pressure on the equilibrium position and the need to consider pressure when constructing a manufacturing plant to give reasons why a pressure of 60–70 atmospheres is used in the reaction between ethene and steam.

   d) Consider the catalyst. The phosphoric(V) catalyst has no effect on the position of equilibrium, so describe why is it needed.

End of topic checklist

Fermentation is the process by which ethanol is made from a solution of sugar and yeast.

Fractional distillation is a process used to separate liquids with different boiling points.

The facts and ideas that you should know and understand by studying this topic:

○ Know that alcohols have a functional group –OH.

○ Know that ethanol is the most common alcohol.

○ Know the structural and displayed formulae for methanol, ethanol, propanol (propan-1-ol only) and butanol (butan-1-ol only).

○ Be able to describe the fermentation process used to manufacture ethanol from sugars, such as glucose, including the importance of the absence of air, an optimum temperature (30 °C) and the enzymes in yeast.

○ Be able to describe the manufacture of ethanol using ethene and steam passed over a catalyst of phosphoric acid, with a temperature of 300 °C and pressure of 60–70 atmospheres.

○ Know that ethanol can be oxidised by:

   ● burning air or oxygen (complete combustion)

   ● reaction with oxygen in the air to form ethanoic acid (microbial oxidation)

   ● by heating with potassium dichromate(VI) in dilute sulfuric acid to form ethanoic acid.

End of topic questions

1. What is the general formula for an alcohol? **(1 mark)**

2. EXTENDED Pentanol is an alcohol with five carbon atoms.

    a) What is the molecular formula of pentanol? **(1 mark)**

    b) Draw the displayed formula for pentanol. **(1 mark)**

3. What are the two main sources of sugar used in the manufacture of ethanol by fermentation? **(2 marks)**

4. In the laboratory fermentation experiment to make ethanol from sugar using yeast, explain the importance of the following:

    a) The reaction temperature is kept between the range 25–30 °C. **(2 marks)**

    b) Oxygen from the air cannot enter the reaction flask. **(2 marks)**

5. In the industrial manufacture of ethanol from ethene:

    a) What is the source of the ethene? **(1 mark)**

    b) What is the function of the phosphoric acid? **(1 mark)**

    c) What conditions of temperature and pressure are used? **(2 marks)**

    d) Write a balanced symbol equation for the reaction. **(1 mark)**

6. Ethanol can be oxidised to ethanoic acid.

    a) Explain why this can be a problem when making ethanol by the fermentation of glucose. **(2 marks)**

    b) In the laboratory the oxidation of ethanol can be achieved using potassium dichromate(VI). What reaction conditions are needed for this reaction? **(2 marks)**

△ Fig. 4.34 Carboxylic acids can be made into esters, which provide the smell and flavour of some sweets.

Carboxylic acids

INTRODUCTION

The first member of the carboxylic acid homologous series is methanoic acid, which ants produce when they sting an animal or attack another insect. The most common carboxylic acid is ethanoic acid, or acetic acid as it used to be called, which is the main constituent of vinegar. As well as having properties similar to those of common acids, such as hydrochloric acid and sulfuric acid, the carboxylic acids have some very different properties.

KNOWLEDGE CHECK

✓ Understand the term homologous series.
✓ Know some of the typical properties of common acids such as hydrochloric acid and sulfuric acid.

LEARNING OBJECTIVES

✓ Know that carboxylic acids contain the functional group $-\overset{\displaystyle O}{\underset{\displaystyle |}{C}}-OH$.
✓ Understand how to draw structural and displayed formulae for unbranched-chain carboxylic acids with up to four carbon atoms in the molecule, and name each compound.
✓ Describe the reactions of aqueous solutions of carboxylic acids with metals and metal carbonates.
✓ Know that vinegar is an aqueous solution containing ethanoic acid.

WHAT ARE CARBOXYLIC ACIDS?

Carboxylic acids are molecules that contain the –COOH functional group, which is responsible for their properties and reactions. The carboxylic acids have the general formula $C_nH_{2n+1}COOH$ (although n can be equal to zero in the case of methanoic acid).

Acid	Formula	Structural formula
Methanoic acid	HCOOH	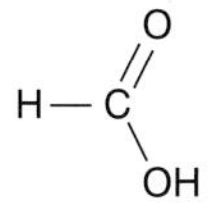
Ethanoic acid	CH_3COOH	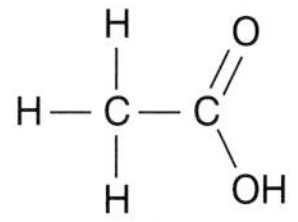
Propanoic acid	C_2H_5COOH	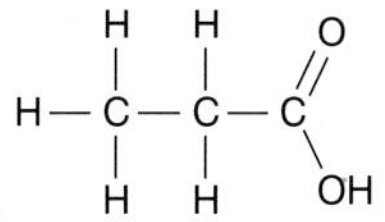
Butanoic acid	C_3H_7COOH	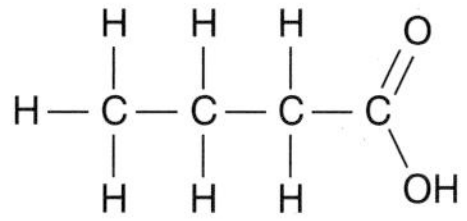

△ Table 4.10 Some common carboxylic acids.

Ethanoic acid is a weak acid because it is only partially ionised when dissolves in water.

$$CH_3COOH(aq) \rightleftharpoons CH_3COO^-(aq) + H^+(aq)$$

THE REACTIONS OF CARBOXYLIC ACIDS

Aqueous ethanoic acid behaves as a typical acid. It has a pH lower than 7 and will react with alkalis and carbonates to form salts. For example, aqueous ethanoic acid will react with sodium carbonate as follows:

ethanoic acid + sodium carbonate ⟶ sodium ethanoate + water + carbon dioxide

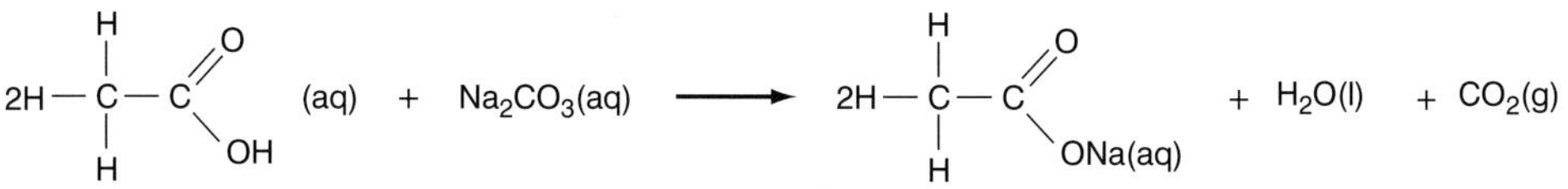

△ Fig. 4.35 The reaction of aqueous ethanoic acid with sodium carbonate.

The reaction will not be as vigorous as with dilute hydrochloric acid of the same concentration. Ethanoic acid is a weak acid and has a lower concentration of hydrogen ions, $H^+(aq)$.

In the same way carboxylic acids will react with metals such as magnesium to produce hydrogen gas.

ethanoic acid + magnesium → magnesium ethanoate + hydrogen

$$2CH_3COOH(aq) + Mg(s) \rightarrow (CH_3COO)_2Mg(aq) + H_2(g)$$

End of topic checklist

The facts and ideas that you should know and understand by studying this topic:

○ Know that carboxylic acids contain the functional group –COOH.

○ Know the names, structural and displayed formulae of methanoic, ethanoic, propanoic and butanoic acids.

○ Be able to describe the reactions of aqueous solutions of carboxylic acids with metals and metal carbonates.

○ Know that vinegar is an aqueous solution containing ethanoic acid.

End of topic questions

1. What is the functional group in a carboxylic acid? **(1 mark)**

2. **a**) What is the name of the carboxylic acid with three carbon atoms in the molecule? **(1 mark)**

   b) Draw the structural formula of this carboxylic acid with three carbon atoms. **(1 mark)**

3. **a**) What would you observe when solid sodium carbonate is added to an aqueous solution of ethanoic acid? **(3 marks)**

   b) Write a balanced equation for this reaction. **(2 marks)**

Esters

INTRODUCTION

Esters are a homologous series of organic compounds made by the reaction between an alcohol and a carboxylic acid. Esters are volatile substances (readily form a vapour), have distinctive smells and are commonly used as food flavourings and in perfumes. The ester grouping is also an important grouping in some synthetic polymers as you will see in the next topic.

△ Fig 4.36 Oils from plants, such as olive oil, are esters of long-chain carboxylic acids and the alcohol propan-1,2,3-triol.

KNOWLEDGE CHECK

✓ Know the functional group of the alcohol homologous series.
✓ Know the functional group of the carboxylic acid homologous series.

LEARNING OBJECTIVES

✓ Know that esters contain the functional group $-\overset{\overset{\displaystyle O}{\|}}{C}-O-$
✓ Know that ethyl ethanoate is the ester produced when ethanol and ethanoic acid react in the presence of an acid catalyst.
✓ Understand how to write the structural and displayed formulae of ethyl ethanoate.
✓ Understand how to write the structural and displayed formulae of an ester given the name or formulae of the alcohol and carboxylic acid from which it is formed and vice versa.
✓ Know that esters are volatile compounds with distinctive smells and are used as food flavourings and perfumes.
✓ Be able to prepare a sample of an ester such as ethyl ethanoate.

WHAT ARE ESTERS?

Esters contain the functional group $-\overset{\overset{\displaystyle O}{\|}}{C}-O-$ and have the general formula $C_nH_{2n+1}COOC_mH_{2m+1}$ where typically m and n are numbers from 1 upwards.

The structural and displayed formulae for ethyl ethanoate are shown in Table 4.11.

Structural formula	Displayed formula
$CH_3COOC_2H_5$	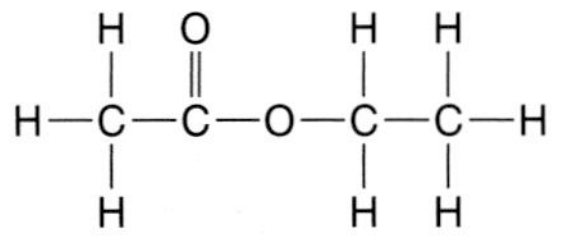

Δ Table 4.11 Structural and displayed formulae of ethyl ethanoate.

Ethyl ethanoate can be made by the reaction between ethanol and ethanoic acid in the presence of concentrated sulfuric acid acting as a catalyst.

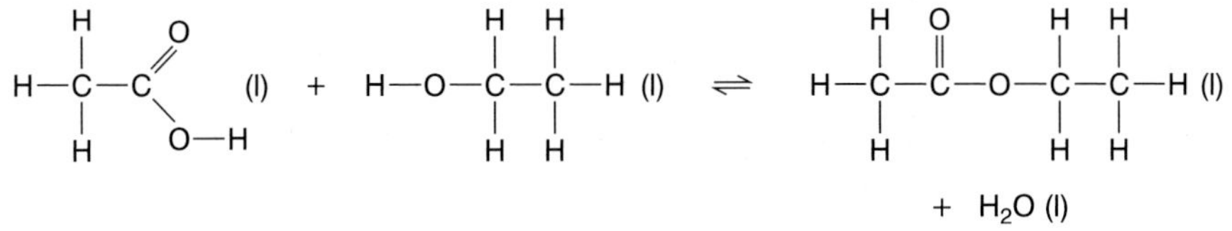

ethanol + ethanoic acid ⇌ ethyl ethanoate + water

Δ Fig.4.37 Reaction of ethanol and ethanoic acid to form an ester, ethyl ethanoate, and water.

Other esters can be made in the same way. For example:

carboxylic acid	+	alcohol	⇌	ester	+	water
propanoic acid	+	propanol	⇌	propyl propanoate	+	water
propanoic acid	+	ethanol	⇌	ethyl propanoate	+	water

The name of the ester identifies which carboxylic acid and which alcohol has been used to make it. For example:

Name of ester	Name of alcohol	Name of carboxylic acid
methyl ethanoate	methanol	ethanoic acid
ethyl methanoate	ethanol	methanoic acid

Esters are sweet-smelling substances. They are commonly used as constituents of perfumes, essential oils, food flavourings (e.g. in pear drop sweets) and in cosmetics.

Developing investigative skills

A sample of ethyl ethanoate was made by mixing some ethanol and ethanoic acid in a boiling tube. A few drops of concentrated sulfuric acid were added and then the boiling tube was warmed up to 60 °C in a water bath heated by a hot plate. After 3 or 4 minutes the contents of the boiling tube were poured into another beaker containing cold water. The smell of the ethyl ethanoate could then be clearly detected coming from the beaker of cold water.

Evaluate data and methods

❶ What safety precautions should you take when using concentrated sulfuric acid?

❷ Why do you think the boiling tube was heated in a beaker of water on a hot plate rather than directly in a Bunsen burner flame?

❸ If you wanted a pure sample of ethyl ethanoate what procedure could be used to separate the ethyl ethanoate that has formed from any unreacted ethanol, ethanoic acid and water?

❹ Adding the reaction mixture to cold water made the ethyl ethanoate smell more apparent rather than that of any ethanoic acid that had not reacted. Can you explain why this approach might work?

End of topic checklist

An **ester** is a compound formed by the reaction between an alcohol and a carboxylic acid in the presence of an acid catalyst.

The facts and ideas that you should know and understand by studying this topic:

○ Know the functional group of the ester homologous series is $-\!\overset{\displaystyle O}{\underset{\displaystyle \|}{C}}\!-\!O\!-$.

○ Know that esters are volatile compounds with distinctive smells and are used in food flavourings and in perfumes.

○ Know the structural and displayed formula for ethyl ethanoate.

○ Be able to describe how ethyl ethanoate can be made from ethanol and ethanoic acid.

○ Understand how to identify which acid and which alcohol are needed to made a particular ester and vice versa.

End of topic questions

1. Draw the displayed formulae for the following esters:

   a) methyl methanoate. (1 mark)

   b) propyl ethanoate. (1 mark)

2. **a)** Copy and complete the following table:

Name of ester	Name of alcohol	Name of carboxylic acid	
ethyl ethanoate			(2 marks)
propyl methanoate			(2 marks)
methyl propanoate			(2 marks)

   b) What catalyst would be needed in each of the above reactions? (1 mark)

Synthetic polymers

INTRODUCTION

Polymers exist in nature; two examples are cellulose and starch. Synthetic polymers are manufactured, often using starting materials made from crude oil. Many synthetic polymers are commonly known as plastics and have a wide range of uses in daily life. Other synthetic polymers are used as fabrics – look for 'polyester' or 'polyamide' next time you look at a label on clothes. Plastics are useful because of their resistance to other chemicals. This is also the reason why disposal of plastic material is such a big environmental issue: they do not break down (degrade) easily.

△ Fig. 4.38 Most tennis racquets have carbon fibre reinforced polymer frames.

KNOWLEDGE CHECK

✓ Know that alkenes are unsaturated hydrocarbons and contain C=C double bonds.
✓ Know that alkenes can be used to make a wide range of plastic materials.
✓ Know how esters are made from carboxylic acids and alcohols.

LEARNING OBJECTIVES

✓ Know that an addition polymer is formed by joining up many small molecules called monomers.
✓ Understand how to draw the repeat unit of an addition polymer, including poly(ethene), poly(propene), poly(chloroethene) and (poly)tetrafluoroethene.
✓ Understand how to deduce the structure of a monomer from the repeat unit of an addition polymer and vice versa.
✓ Be able to explain problems in the disposal of addition polymers, including:
  • their inertness and inability to biodegrade
  • the production of toxic gases when they are burned.
✓ Know that condensation polymerisation, in which a dicarboxylic acid reacts with a diol, produces a polyester and water.
✓ Understand how to write the structural and displayed formula of a polyester, showing the repeat unit, given the formulae of the monomers from which it is formed including the reaction of ethanedioic acid and ethanediol.
✓ Know that some polyesters, known as biopolyesters, are biodegradable.

ADDITION POLYMERISATION

Alkenes can be used to make **polymers**, which are very large molecules made up of many identical smaller molecules called **monomers**. Alkenes are able to react with themselves. They join together into long chains, like adding beads to a necklace. When the monomers add together like this, the material produced is called an **addition polymer**. Poly(ethene) or polythene is made this way.

By changing the atoms or groups of atoms attached to the carbon–carbon double bond, a whole range of different polymers can be made.

The double bond within the alkene molecule breaks to form a single covalent bond to a carbon atom in an adjacent molecule. This process is repeated rapidly as the molecules link together.

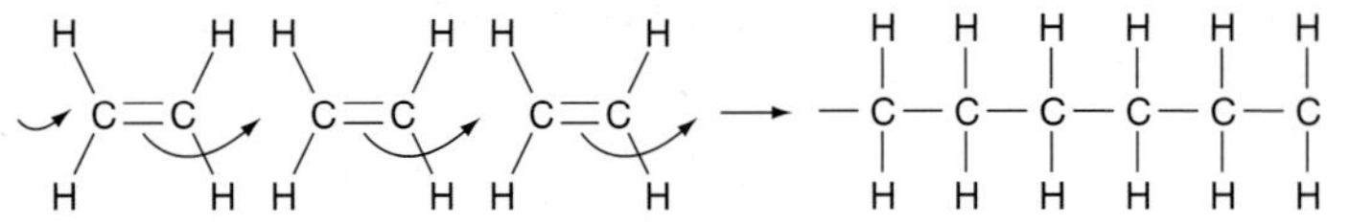

△ Fig. 4.40 Poly(ethene) from ethene

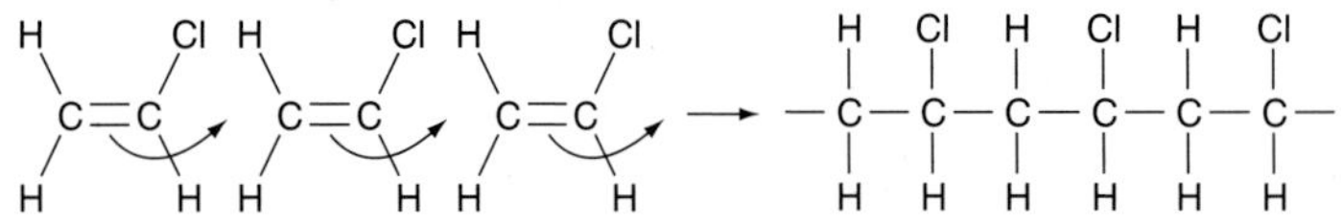

△ Fig. 4.41 Poly(chloroethene) from chloroethene

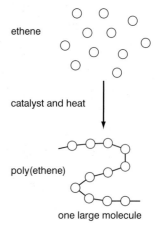

△ Fig. 4.39 Ethene molecules link together to produce a long polymer chain of poly(ethene).

Name of monomer	Displayed formula of monomer	Name of polymer	Displayed formula of polymer	Uses of polymer
ethene	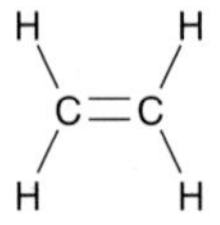 H₂C=CH₂	poly(ethene)	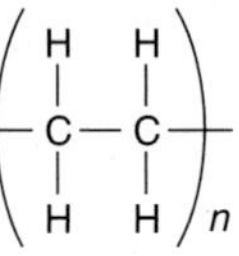 $\left(-CH_2-CH_2-\right)_n$	buckets, bowls, plastic bags
propene	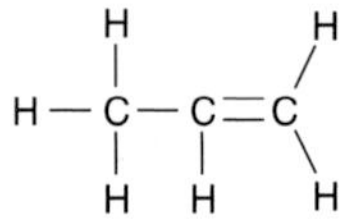 H₃C-CH=CH₂	poly(propene)	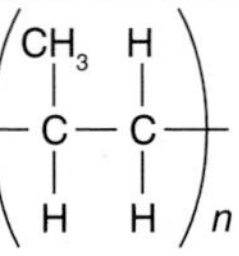 $\left(-CH(CH_3)-CH_2-\right)_n$	packaging, ropes, carpets
chloroethene (vinyl chloride)	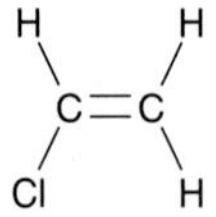 CHCl=CH₂	poly(chloroethene) (polyvinylchloride)	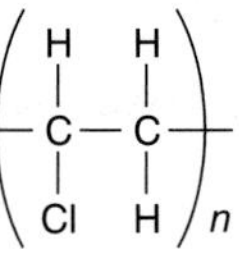 $\left(-CHCl-CH_2-\right)_n$	plastic sheets, artificial leather
tetrafluoroethene	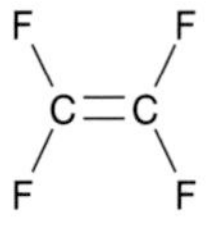 F₂C=CF₂	poly(tetrafluroethene) or PTFE	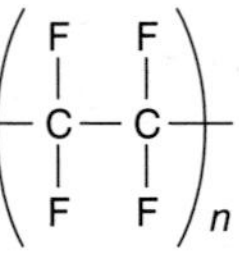 $\left(-CF_2-CF_2-\right)_n$	non-stick coating in frying pans

△ Table 4.12 Monomers and their polymers.

◁ Fig. 4.42 This pan is coated with PTFE non-stick plastic. How is PTFE made?

QUESTIONS

1. In what way is a polymer like a string of beads?

2. Name the polymer used extensively to make plastic bags.

3. The diagram shows the structural formula of chloroethene.

a) Show how two molecules of chloroethene join together to form part of the polymer poly(chloroethene).

b) Draw the structure of the repeat unit of the polymer.

4. What type of polymer is poly(chloroethene)?

CONDENSATION POLYMERISATION

Polymers can also be made by joining together two different monomers so that they react together. When they react, they expel a small molecule. Because the molecule is usually water, the process is called condensation **polymerisation** and the products are **condensation polymers**.

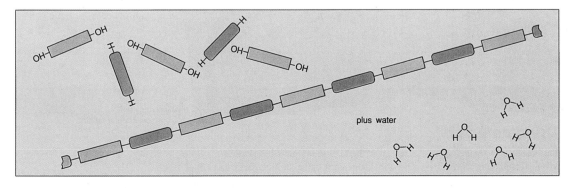

plus water

△ Fig. 4.43 Condensation polymerisation.

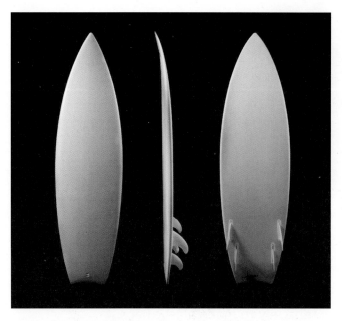

△ Fig. 4.44 This surfboard is made from a condensation polymer called polyurethane.

Polyesters are condensation polymers and, as the name suggests, just like esters they are made from an alcohol and a carboxylic acid. However, so that ester links can be formed between a large number of carboxylic acid and alcohol monomers each monomer needs to have two functional groups. The alcohol is known as a **diol** and the acid as a **dicarboxylic acid**. How these combine together is shown in the diagram below.

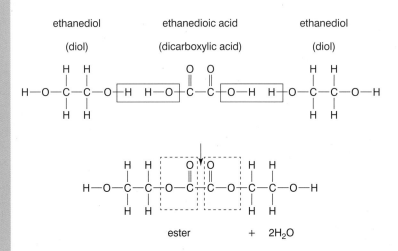

△ Fig.4.45 Reaction of a diol and a dicarboxylic acid to produce an ester linkage and water.

QUESTIONS

1. Polyester is an example of a condensation polymer. How is this type of polymer different from an addition polymer?

2. Why is it important that the monomers which join together to make polyester have reactive groups at each end of the molecule?

Many of the plastics in common use are addition polymers. The pollution caused by plastics is a growing problem, particularly as plastics are durable and inexpensive. Many plastics are non-biodegradable and so cannot be broken down naturally by bacteria in the environment. Living organisms, particularly marine animals, can be affected through entanglement, direct ingestion of plastic waste or through exposure to chemicals within the plastics. In the UK alone 5 million tonnes of plastic are consumed each year, and it is estimated that less than one-quarter of this enters recycling systems. This leaves large amounts of plastic waste destined for burial in landfill sites or for burning. Burning plastics generates highly toxic fumes.

Recycling is also not without significant challenges. Different plastics have to be carefully separated to identify those that can be recycled (see the section on thermoplastics and thermosetting plastics).

Recently bioplastics and biopolyesters, made from renewable biomass sources, such as vegetable fats and oils and starch, have been developed and a number of these are biodegradable.

◁ Fig. 4.46 Some plastic bags are biodegradable. They are made from polythene and starch. When buried, the starch breaks down and leaves tiny fragments of polythene.

SCIENCE IN CONTEXT

THE CHALLENGES OF RECYCLING PLASTICS

△ Fig. 4.47 Waste plastics are lightweight but very bulky.

Some types of plastic can be melted down and used again. These are called thermoplastics. Other types of plastic harden or decompose when they are heated. These are called thermosetting plastics. Recycling them is difficult because the different types of plastic must be separated.

▷ Fig. 4.48 Thermoplastics have weak intermolecular forces that break on heating, which allows them to be melted and re-moulded. In thermosetting plastics, the intermolecular bonds are strong interlinking covalent bonds. The whole structure eventually breaks down when these bonds are broken by heating.

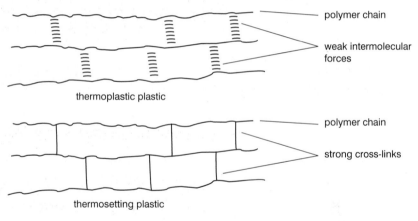

End of topic checklist

A **monomer** is a molecule that can combine with other molecules to form a **polymer.**

An **addition polymer** is made when molecules of a single monomer join together in large numbers.

A **condensation polymer** is formed when two monomers react together and eliminate a small molecule such as water

Biopolyesters are biodegradable.

The facts and ideas that you should know and understand by studying this topic:

○ Understand that an addition polymer is formed by joining up many small molecules called monomers.

○ Be able to draw repeat units of addition polymers from the structural formula of the monomer, such as poly(ethene), poly(propene), poly(chloroethene) and poly(tetrafluroethene).

○ Be able to deduce the structural formula of a monomer from the repeat unit of an addition polymer, and do the same in reverse.

○ Be able to explain that addition polymers are hard to dispose of as their inertness means that they do not easily biodegrade, and they produce toxic gases when burnt.

○ Understand that some polymers, such as polyester, form by a different process called condensation polymerisation.

○ Know that when dicarboxylic acid reacts with a diol, this produces a polyester and water and so is condensation polymerisation.

○ Given the formulae of the monomers from which it is formed, be able to draw the structural and displayed formula of a polyester showing the repeat unit, including the reaction of ethanedioic acid and ethanediol.

End of topic questions

1. Explain what is meant by the following words:

   a) monomer   **(1 mark)**

   b) polymer.   **(1 mark)**

2. This question is about addition polymers.

   a) What is an addition polymer?   **(1 mark)**

   b) What structural feature do all monomers that form addition polymers have in common?   **(1 mark)**

   c) Use displayed formulae to show how propene molecules react together to form poly(propene).   **(2 marks)**

   d) Explain why a polymer poly(propane) does not exist.   **(2 marks)**

3. Addition polymers are not biodegradable. Explain what this means.   **(2 marks)**

4. This question is about condensation polymers.

   a) In what ways is a condensation polymer different to an addition polymer?   **(2 marks)**

   b) Name an example of a condensation polymer.   **(1 mark)**

   c) Show how a polyester is formed by the reaction between a diol and a dicarboxylic acid.   **(1 mark)**

5. A headline in a newspaper states 'Plastics – the problem must be solved'. Discuss the problem referred to and how you think it should be solved.   **(6 marks)**

Exam-style questions
Sample student answer

Question 1

This question is about the following organic compounds:

A C_4H_{10} **B** C_2H_5OH **C** C_4H_8 **D** CH_3COOH

a) Which compound belongs to the alkene group?

C ✓   (1)

b) Which compound is a saturated hydrocarbon?

C ✗   (1)

c) Which compound can be made by fermentation?

B ✓   (1)

d) Which compound will decolourise bromine water?

C ✓   (1)

e) Draw the structural and displayed formula for compound A.

structural : $CH_3 \ CH_2 \ CH_2 \ CH_3$

displayed :

```
      H   H   H   H
      |   |   |   |
  H — C — C — C — C — H
      |   |   |   |
      H   H   H   H
```

butane   (2)

f) Name the two products that are formed when compound A burns in a plentiful supply of air.

carbon dioxide ✓ *and water* ✓   (2)

g) Name two other substances that can be produced when compound A burns in a limited supply of air.

carbon ✓ *and hydrogen* ✗   (2)

Total   9/10

EXAMINER'S COMMENTS

a) Correct answer, alkenes are hydrocarbons with the general formula C_nH_{2n}.

b) The correct answer is A. Compound C is an alkene and is unsaturated. It contains a carbon–carbon double bond.

c) Correct answer, ethanol is made by fermentation.

d) Correct answer. The decolourising of bromine water is a test for an alkene.

e) Correct answers.

f) Correct answers. All hydrocarbons burn in a plentiful supply of oxygen forming carbon dioxide and water.

g) Carbon is correct and is responsible for the yellow flame characteristic of incomplete combustion. The missing substance is carbon monoxide, which is extremely poisonous.

Question 2

Many useful substances are produced by the fractional distillation of crude oil.

a) Bitumen, fuel oil and gasoline are three fractions obtained from crude oil.

  Name the fraction that has the following property:

  i) the highest boiling point .... (1)

  ii) molecules with the fewest carbon atoms .... (1)

  iii) the darkest colour .... (1)

b) Some long-chain hydrocarbons can be broken down into more useful
   products. What is the name of this process and how is it carried out? .... (3)

c) Methane is used as a fuel. When methane is burned in a limited supply of air carbon
   monoxide is formed.

  i) Write a balanced symbol equation for this reaction. .... (2)

  ii) Explain why carbon monoxide is dangerous to health. .... (2)

(**Total 10 marks**)

Question 3

The alkanes are a homologous series of saturated hydrocarbons.

a) Say whether each of the following statements about the members of the alkane
   homologous series is TRUE or FALSE.

  i) They have similar chemical properties.

  ii) They have the same displayed formula.

  iii) They have the same general formula.

  iv) They have the same physical properties.

  v) They have the same relative formula mass. .... (2)

b) Define the following terms:

  i) hydrocarbon .... (1)

  ii) saturated .... (1)

c) The third member of the alkane homologous series is propane.

  i) What is the molecular formula of propane? .... (1)

  ii) Draw the displayed formula of propane. .... (2)

(**Total 7 marks**)

Exam-style questions cont.

Question 4

The alkenes are a homologous series of unsaturated hydrocarbons. The table below gives the names, formulae and boiling points of the first few members of the series.

Name	Formula	Boiling point (°C)
ethene	C_2H_4	−102
propene	C_3H_6	−48
butene	C_4H_8	−7
pentene	C_5H_{10}	30
hexene		

a) What are the molecular formula, structural formula and displayed formula of hexene? (3)

b) Will hexene be a liquid or gas? Explain your answer. (2)

c) Predict the boiling point of hexene. (1)

d) Describe a test you could use to distinguish between hexene and hexane.

   i) Test:

   ii) Result with hexene:

   iii) Result with hexane: (3)

e) Draw the displayed formulae of the two isomers of butene. (2)

f) Ethene reacts with steam in the presence of a catalyst.

   i) Name the product of the reaction. (1)

   ii) Name a catalyst that could be used in the reaction. (1)

   iii) Write a balanced symbol equation with state symbols for the reaction. (2)

(Total 15 marks)

Exam-style questions cont.

Question 5

Sugar can be converted into ethene in the two-stage process shown below.

	Reaction 1		Reaction 2	
Sugar	$\rightarrow$	Ethanol	$\rightarrow$	Ethene

a) State the type of reaction involved in:

   i) Reaction 1 .... (1)

   ii) Reaction 2 .... (1)

b) State two reaction conditions needed in Reaction 1.

   i) Condition 1: .... (1)

   ii) Condition 2: .... (1)

 c) Write the chemical equation for Reaction 2. .... (1)

d) Ethanol is also manufactured from ethene. Give three conditions for this reaction.

   i) Condition 1: .... (1)

   ii) Condition 2: .... (1)

   iii) Condition 3: .... (1)

 e) Draw the displayed formula for ethanol. .... (1)

(Total 9 marks)

Exam-style questions cont.

Question 6

Propene can be converted into a polymer called poly(propene).

propene

a) Which homologous series does propene belong to? .... **(1)**

b) What is the general name given to an individual molecule that combines with other molecules to make a polymer? .... **(1)**

c) Use the displayed formula of propene to show how three molecules link together to form part of the poly(propene) molecule. .... **(2)**

d) Draw the repeat unit in poly(propene). .... **(2)**

e) What type of polymer is poly(propene)? .... **(1)**

f) Why are polymers such as poly(propene) hard to dispose of? .... **(2)**

g) Polyester is a different type of polymer to poly(propene).

   i) What type of polymer is polyester? .... **(1)**

   ii) In terms of how it is made, how is polyester different to poly(propene)? .... **(1)**

h) i) Draw the displayed formulae for ethanedioic acid .... **(1)**

   ii) Draw the displayed formulae for ethanediol .... **(1)**

   iii) Show how ethanedioic acid and ethanediol can form an ester linkage. **(2)**

(Total 15 marks)

The International GCSE examination

INTRODUCTION

The International GCSE examination tests how good your understanding of scientific ideas is, how well you can apply your understanding to new situations and how well you can analyse and interpret information you have been given. The assessments are opportunities to show how well you can do these.

To be successful in exams you need to:

✓ have a good knowledge and understanding of science

✓ be able to apply this knowledge and understanding to familiar and new situations

✓ be able to interpret and evaluate evidence that you have just been given.

You need to be able to do these things under exam conditions.

OVERVIEW

The International GCSE course is designed to provide a basis for progression to further study in GCE Advanced Subsidiary Chemistry, Advanced Level Chemistry and the International Baccalaureate.
The relationship of the assessment to the qualifications available is shown below:

Biology Paper 1   +   Biology Paper 2   →   International GCSE in Biology

2 hours   1 hour

+

Chemistry Paper 1   + **Chemistry Paper 2**   → **International GCSE in Chemistry**

2 hours   **1 hour 15 minutes**

+

Physics Paper 1   +   Physics Paper 2   →   International GCSE in Physics

2 hours   1 hour

↓

International GCSE in Science (Double Award)

Paper 1 is marked out of 110 and contributes 61.1% of the total International GCSE marks.

Paper 2 is marked out of 70 and contributes 38.9% of the International GCSE marks.

On the Edexcel Double Award Science course Paper 1 accounts for 33.3% of the overall marks.

There is no separate assessment of investigative skills – the assessment is included in the two written papers.

There will be a range of short-answer structured questions, along with some multiple choice questions and a few questions requiring longer answers in both papers. You will be required to perform calculations, draw graphs and describe, explain and interpret chemical ideas and information. In some of the questions the content may be unfamiliar to you; these questions are designed to assess data-handling skills and the ability to apply chemical principles and ideas in unfamiliar situations.

ASSESSMENT OBJECTIVES AND WEIGHTINGS

The assessment objectives and weightings are as follows:

✓ AO1: Knowledge and understanding (38–42%)

✓ AO2: Application of knowledge and understanding, analysis and evaluation (38–42%)

✓ AO3: Experimental skills, analysis and evaluation of data and methods (19–21%).

The types of questions in your assessment fit the three assessment objectives shown in the table.

Assessment objective	Your answer should show that you can...
AO1 recall the science	Recall, select and communicate your knowledge and understanding of science.
AO2 apply your skills and knowledge	Apply skills, including evaluation and analysis, knowledge and understanding of scientific contexts.
AO3 use experimental skills	Use the skills of planning, observation, analysis and evaluation in practical situations.

EXAMINATION TIPS

To help you get the best results in exams, there are a few simple steps to follow.

Check your understanding of the question

✓ **Read the introduction to each question carefully before moving on to the questions themselves**.

✓ Look in detail at any **diagrams, graphs** or **tables**.

✓ Underline or circle the **key words** in the question.

✓ **Make sure you answer the question that is being asked** rather than the one you wish had been asked!

REMEMBER

Remember that any information you are given is there to help you to answer the question.

Make sure that you understand the meaning of the '**command words**' in the questions.

EXAMPLE

- '**Give**', '**state**', '**name**' are used when recall of knowledge is required, for example you could be asked to give a definition or make a list of examples.

- '**State what is meant by**' is used when the meaning of a term is expected but there are different ways for how these can be described.

- '**Identify**' is usually used when you have to select some key information from a text or diagram in the question.

- '**Describe**' is used when you have to give the main feature(s) of, for example, a chemical process or structure. You do not need to include a justification or reason.

- '**Explain**' is used when you have to give reasons, e.g. for some experimental results or a chemical fact or observation. You will often be asked to '**justify**' or 'explain your answer', i.e. give reasons for it.

- '**Suggest**' is used when you have to come up with an idea to explain the information you're given – there may be more than one possible answer, no definitive answer from the information given, or it may be that you will not have learnt the answer but have to use the knowledge you do have to come up with a sensible one.

- '**Calculate**' means that you have to work out an answer in figures, showing relevant working.

- '**Determine**' means you must use data from the question, or must show how the answer can be reached quantitatively. To gain maximum marks, there must be a quantitative element to the answer.

- **'Estimate'** means find an approximate value, number or quantity from a diagram/given data or through a calculation.

- **'Plot'**, **'Draw a graph'** are used when you have to use the data provided to produce graphs and charts. This includes drawing a line of best fit through the points you have plotted. A suitable scale and appropriately labelled axes must be included if these are not provided in the question.

- **'Sketch'** means produce a drawing by hand. For a graph, this would need a line and labelled axes with important features indicated. The axes are not scaled.

- **'Draw'** means produce a diagram either using a ruler or by hand.

- **'Add/Label'** is used when you have to add or label something given in the question, for example, labelling a diagram or adding units to a table.

- **'Complete'** means complete a table/diagram given in the question.

- **'Comment on'** is used when you have to bring together a number of variables from data/information to form a judgement.

- **'Deduce'** means draw/reach conclusion(s) from the information provided.

- **'Discuss'** is used when you have to:
  - Identify the issue/situation/problem/argument that is being assessed within the question.
  - Explore all aspects of an issue/situation/problem/argument.
  - Investigate the issue/situation etc. by reasoning or argument.

- **'Evaluate'** means review information (e.g. data, methods) then bring it together to form a conclusion, drawing on evidence including strengths, weaknesses, alternative actions, relevant data or information. Come to a supported judgement of a subject's quality and relate it to its context.

- **'Give a reason/reasons'** is used when a statement has been made and you only have to give the reason(s) why.

- **'Justify'** means give evidence to support (either the statement given in the question or an earlier answer).

- **'Design'** means plan or invent a procedure from existing principles/ideas.

- **'Predict'** Give an expected result.

- **'Show that'** means verify the statement given in the question.

Check the number of marks for each question

✓ Look at the **number of marks** allocated to each question.

✓ Make sure you include at least as many points in your answer as there are marks.

✓ Look at the space provided to guide you as to the length of your answer. However, be aware that there may be more space than you need.

What to do if you need extra space to answer

✓ If you need more space to answer than provided, then either use the nearest available space, e.g. at the bottom of the page, or ask for extra paper.

✓ However, do NOT use any blank pages in the examination paper as these will not be scanned for the Examiner.

✓ You MUST state clearly WITHIN the marked-out answer space where you have continued your answer, e.g. 'continued at the bottom of page 12', otherwise your additional material may not be seen by the Examiner.

✓ You MUST make it clear IN YOUR ANSWER which question you are answering.

✓ If you use extra paper you MUST make sure you also include your name, candidate number and centre number.

REMEMBER

Beware of continually writing too much because it probably means you are not really answering the questions.

Use your time effectively

✓ Don't spend so long on some questions that you don't have time to finish the paper.

✓ You should spend approximately **one minute per mark**.

✓ If you are really stuck on a question, leave it, finish the rest of the paper and come back to it at the end.

✓ Even if you eventually have to guess at an answer, you stand a better chance of gaining some marks than if you leave it blank.

ANSWERING QUESTIONS

Multiple choice questions

✓ Select your answer by placing a cross (not a tick) in the box.

Short and long answer questions

✓ In short-answer questions, **don't write more than you are asked for**.

✓ You will not gain any marks, even if the first part of your answer is correct, if you've written down something incorrect later on or which contradicts what you've said earlier. This just shows that you haven't really understood the question or are guessing.

- ✓ In some questions, particularly short-answer questions, answers of only one or two words may be sufficient, but in longer questions you should aim to use **good English** and **scientific language** to make your answer as clear as possible.
- ✓ Present the information in a logical sequence.
- ✓ Don't be afraid to use **labelled diagrams** or **flow charts** if it helps you to show your answer more clearly.

Questions with calculations

- ✓ **In calculations always show your working.**
- ✓ Even if your final answer is incorrect you may still gain some marks if part of your attempt is correct.
- ✓ If you just write down the final answer and it is incorrect, you will get no marks at all.
- ✓ Write down your answers to as many **significant figures** as are used in the numbers in the question (and no more). If the question doesn't state how many significant figures then a good rule of thumb is to quote 3 significant figures.
- ✓ Don't round off too early in calculations with many steps – it's always better to give too many significant figures than too few.
- ✓ You may also lose marks if you don't use the correct **units.** In some questions the units will be mentioned, e.g. calculate the mass in grams; or the units may also be given on the answer line. If numbers you are working with are very large, you may need to make a conversion, e.g. convert joules into kilojoules, or kilograms into tonnes.

Finishing your exam

- ✓ When you've finished your exam, **check through** your paper to make sure you've answered all the questions.
- ✓ Check that you haven't missed any questions at the end of the paper or turned over two pages at once and missed questions.
- ✓ Cover over your answers and read through the questions again and check that your answers are as good as you can make them.

REMEMBER

In the two written papers, you will be asked questions on investigative work (Assessment objectives AO3 and AO2). It is important that you understand the methods used by scientists when carrying out investigative work.

More information on carrying out practical work and developing your investigative skills are given in the next section.

A copy of the Periodic Table will be provided at the front of the examination papers.

Developing experimental skills

INTRODUCTION

As part your International GCSE Chemistry course, you will develop practical skills and have to carry out investigative work in science.

This section provides guidance on carrying out an investigation.

Many investigations follow a common route:

1. Planning the investigation and assessing the risk

2. Carrying out the practical work safely and skilfully

3. Making and recording observations and measurements

4. Analysing the data and drawing conclusions

5. Evaluating the data and methods used

1. Planning and assessing the risk

Learning objective: to devise and plan investigations, drawing on chemical knowledge and understanding in selecting appropriate techniques.

Questions to ask

What do I already know about the area of chemistry I am investigating and how can I use this knowledge and understanding to help me with my plan?

✓ Think about what you have already learned and any investigations you have already done that are relevant to this investigation.

✓ List the factors might affect the process you are investigating.

What is the best method or technique to use?

✓ Think about whether you can use or adapt a method that you have already used.

✓ A method, and the measuring instruments, must be able to produce **valid** measurements. A measurement is valid if it measures what it is supposed to be measuring.

You will make a decision as to which technique to use based on:

✓ The accuracy and precision of the results required.

Investigators might require results that are as accurate and precise as possible but if you are doing a quick comparison, or a preliminary test to check a range over which results should be collected, a high level of accuracy and precision may not be required.

✓ The simplicity or difficulty of the techniques available, or the equipment required; is this expensive, for instance?

✓ The scale, e.g. using standard laboratory equipment or on a micro-scale, which may give results in a shorter time period.

✓ The time available to do the investigation.

✓ Health and safety considerations.

What am I going to measure?

✓ The factor you are investigating is called the **independent variable**. A **dependent variable** is affected or changed by the independent variable that you select.

✓ You need to choose a range of measurements that will be enough to allow you to plot a graph of your results and so find out the pattern in your results.

✓ You might be asked to explain why you have chosen your range rather than a lower or higher range.

How am I going to control the other variables?

✓ These are **control variables**. Some of these may be difficult to control.

✓ You must decide how you are going to control any other variables in the investigation and so ensure that you are carrying out a fair test and that any conclusions you draw are valid.

✓ You may also need to decide on the concentration or combination of reactants.

What equipment is suitable and will give me the accuracy and precision I need?

✓ The **accuracy** of a measurement is how close it is to its true value.

✓ **Precision** is related to the smallest scale division on the measuring instrument that you are using, e.g. when measuring a distance, a rule marked in millimetres will give greater precision that one divided into centimetres only.

✓ A set of precise measurements also refers to measurements that have very little spread about the mean value.

✓ You need to be sensible about selecting your devices and make a judgement about the degree of precision. Think about what is the least precise variable you are measuring and choose suitable measuring devices. There is no point having instruments that are much more precise than the precision you can measure the variable to.

What are the potential hazards of the equipment, chemicals and technique I will be using and how can I reduce the risks associated with these hazards?

✓ You can find out the hazard associated with the two reactants using CLEAPSS Student Safety Sheets or a similar resource.

✓ In the exam, be prepared to suggest safety precautions when presented with details of a chemistry investigation.

EXAMPLE 1

You have been asked to design and plan an investigation to find out the effect of temperature on the rate of reaction between sodium thiosulfate and hydrochloric acid. In a previous investigation you have used this reaction so you are familiar with what happens and how the rate of the reaction can be measured.

What do I already know?

Previously you have used the reaction between sodium thiosulfate and hydrochloric acid to investigate the effect of changing the concentration of sodium thiosulfate on the rate of the reaction. So you know that as the reaction takes place a precipitate of sulfur forms in the solution and makes the solution change from colourless (and clear) to pale yellow (and opaque). The time it takes for a certain amount of sulfur to form can be used as a measure of the rate of the reaction.

What is the best method or technique to use?

The technique you used in your previous investigation can be adapted. Previously you added 5 cm³ of hydrochloric acid to 50cm³ of sodium thiosulfate solution in a conical flask and then looked down the conical flask and measured the time that was taken until a mark on a piece of paper under the flask was no longer visible.

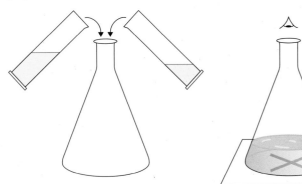

△ Fig. 6.1 Apparatus for experiment.

What am I going to measure?

You are investigating the effect of temperature on the rate of the reaction. The independent variable is temperature. The time it takes for the mark to become obscured by the sulfur forming in the flask is a dependent variable as it depends on the temperature you select.

Other independent variables you could measure are the temperature of the sodium thiosulfate or the mixture once the acid has been added.

You will need to be able to measure a **range** of temperatures (e.g. 10 °C to 50 °C). Ideally you will need to repeat the experiment at about five different temperatures (e.g. 10 °C, 20 °C, 30 °C, 40 °C, and 50 °C).

How am I going to control the other variables?

It is important that you decide on the quantities of sodium thiosulfate and hydrochloric acid (the volumes) you are going to use and then keep these unchanged throughout. As you are familiar with the reaction you can look back at your previous results and decide which concentration or combination of reactants would be the most appropriate.

What equipment is suitable and will give me the accuracy and precision I need?

You now know what you will need to measure and so can decide on your measuring devices.

Measurement	Quantity	Device
Volume (sodium thiosulfate)	about 50 cm³	100 cm³ measuring cylinder
Volume (hydrochloric acid)	about 5 cm³	10 cm³ measuring cylinder
Temperature	10–50 °C	Thermometer (1 °C precision) or a thermostatically controlled water bath
Time	up to 120 s	Stop watch or stopclock (1 s precision)

△ Table 6.1 Suitable equipment for experiment.

Choosing a burette accurate to 0.1 cm³, a thermometer accurate to 0.1 °C or a stop watch accurate to 0.01 s would be inappropriate when the technique itself does not have a very precise way of judging when the mark has been obscured.

What are the potential hazards and how can I reduce the risks?

The hazards are as follows:

✓ Sodium thiosulfate solution: LOW HAZARD

✓ Dilute hydrochloric acid (less than 2.7 mol/dm³: LOW HAZARD

Sulfur dioxide is produced in the reaction. The gas is toxic and asthmatics are particularly at risk. An alternative process involving micro quantities is available (see CLEAPSS guidance).

In terms of the equipment, the major hazard will be handling the hot solution. You can limit this hazard by choosing a range of temperatures that do not include very high values (e.g. between 10 °C and about 50 °C).

2. Carrying out the practical work safely and skilfully

Learning objective: To demonstrate and describe appropriate experimental and investigative methods, including safe and skilful practical techniques.

Questions to ask:

How shall I use the equipment and chemicals safely to minimise the risks – what are my safety precautions?

✓ When writing a Risk Assessment, investigators need to be careful to check that they've matched the hazard with the concentration of a chemical used. Many acids, for instance, are corrosive in higher concentrations, but are likely to be irritants or of low hazard in the concentration used when working in chemistry experiments.

✓ Don't forget to consider the hazards associated with all the chemicals, including the products of the reactions, even if these are very low.

✓ In the exam, you may be asked to justify the precautions taken when carrying out an investigation.

How much detail should I give in my description?

✓ You need to give enough detail so that someone else who has not done the experiment would be able to carry it out to reproduce your results.

How should I use the equipment to give me the precision I need?

✓ You should know how to read the scales on the measuring equipment you are using.

✓ You need to show that you are aware of the precision needed.

△ Fig. 6.2 The volume of liquid in a burette must be read to the bottom of the meniscus. The volume in this measuring cylinder is 202 cm³ (ml), not 204 cm³.

EXAMPLE 2

This is an extract from a student's notebook. It describes how she carried out a titration experiment using sulfuric acid and potassium hydroxide, with methyl orange as an indicator.

$2KOH(aq) + H_2SO_4(aq) \rightarrow K_2SO_4(aq) + 2H_2O(l)$

Safety precautions

a) *Equipment*

I will be using a pipette, burette and conical flask all made from glass so I will need to handle them carefully and, in particular, clamp the burette carefully and make sure the pipette does not roll off the bench when I am not using it. I will also be careful when attaching the pipette to the pipette filler so that the pipette does not break.

b) *Chemicals*

I have looked up the hazards

Sulfuric acid 0.1M: Sulfuric acid at this concentration is not currently classified as hazardous.

Potassium hydroxide (0.1M approx.): IRRITANT

COMMENT

The student has used a data source to look up the chemical hazards. However, there is no risk assessment for methyl orange.

I will need to handle the chemicals carefully, have a damp cloth ready to wipe up any spills and wear eye protection.

COMMENT

The student has suggested some sensible precautions.

The student's method is given below

1. *A pipette and a burette were carefully washed, making sure they were drained after washing.*

2. *25.00 cm³ of the potassium hydroxide solution was measured in a pipette and transferred to a clean conical flask. 3 drops of methyl orange indicator were then added.*

3. *The burette was filled with the sulfuric acid solution, making sure that there were no air bubbles in the jet of the burette. The first reading of the volume of acid in the burette was taken.*

4. *The acid was added to the alkali in the conical flask, swirling the flask all the time. When the indicator colour was close to changing, the acid was added drop by drop until the colour changed. The second reading on the burette was taken.*

5. *The whole procedure was repeated twice, making sure that between each experiment the conical flask was washed carefully. The results are shown in the table.*

COMMENT

The method is well written and detailed. Point 1 could have been improved if she had said that the burette and pipette had been washed with distilled water first and then the chemical to be used (burette – acid; pipette – alkali).

Precision and accuracy. Some examples from the notebook are:

'25.00 cm³ of the potassium hydroxide solution'

The student has appreciated the accuracy a titration can achieve.

'making sure that there were no air bubbles in the jet of the burette'

An air bubble could easily lead to an inaccurate measurement.

'the acid was added drop by drop until the colour changed'

Again the student tried to get accuracy to within one drop (± 0.05 cm³)

3. Making and recording observations and measurements

Learning objective: to make observations and measurements with appropriate precision, record these methodically, and present them in a suitable form.

Questions to ask:

How many different measurements or observations do I need to take?

✓ Sufficient readings have been taken to ensure that the data are consistent.

✓ It is usual to repeat an experiment to get more than one measurement. If an investigator takes just one measurement, this may not be typical of what would normally happen when the experiment was carried out.

✓ When repeat readings are consistent they are said to be **repeatable**.

Do I need to repeat any measurements or observations that are anomalous?

✓ An **anomalous result** or **outlier** is a result that is not consistent with other results.

✓ You want to be sure a single result is accurate (as in the example below). So you will need to repeat the experiment until you get close agreement in the results you obtain.

✓ If an investigator has made repeat measurements, they would normally use these to calculate the arithmetical mean (or just mean or average) of these data to give a more accurate result. You calculate the mean by adding together all the measurements, and dividing by the number of measurements. Be careful though, anomalous results should not be included when taking averages.

✓ Anomalous results might be the consequence of an error made in measurement. But sometimes outliers are genuine results. If you think an outlier has been introduced by careless practical work, you should omit it when calculating the mean. But you should examine possible reasons carefully before just leaving it out.

✓ You are taking a number of readings in order to see a changing pattern. For example, measuring the volume of gas produced in a reaction every 10 seconds for 2 minutes (so 12 different readings). It is likely that you will plot your results onto a graph and then draw a **line of best fit**.

✓ You can often pick an anomalous reading out from a results table (or a graph if all the data points have been plotted, as well as the mean, to show the range of data). It may be a good idea to repeat this part of the practical again, but it's not necessary if the results show good consistency.

✓ If you are confident that you can draw a line of best fit through most of the points, it is not necessary to repeat any measurements that are obviously inaccurate. If, however, the pattern is not clear enough to draw a graph then readings will need to be repeated.

How should I record my measurements or observations – is a table the best way? What headings and units should I use?

✓ A table is often the best way to record results.

✓ Headings should be clear.

- ✓ If a table contains numerical data, do not forget to include units; data are meaningless without them.
- ✓ The units should be the same as those that are on the measuring equipment you are using.
- ✓ Sometimes you are recording observations that are not quantities, as shown in Example 6. Putting observations in a table with headings is a good way of presenting this information.

EXAMPLE 3

The student from Example 2 has recorded the results in a table as shown below. In this case she needs to get two results within 0.1 cm^3 of each other and so has had to do the titration three times to get that level of agreement.

Volume of potassium hydroxide solution = 25.00 cm^3

Burette reading	1st experiment	2nd experiment	3rd experiment
2nd reading (cm^3)	17.50	19.50	20.50
1st reading (cm^3)	0.00	2.50	3.50
Difference (cm^3)	17.50	17.00	17.00

△ Table 6.2 Readings from titration.

EXAMPLE 4

In an experiment to measure the volume of a gas produced in a reaction the student has sensibly recorded her results in a table. Notice each column has a heading *and* units.

Time (s)	Volume of gas (cm^3)
0	25
10	45
20	60
30	70
40	74
50	76
60	76

EXAMPLE 5

In another experiment the student has recorded his results obtained in an experiment involving heating magnesium in a crucible to form magnesium oxide.

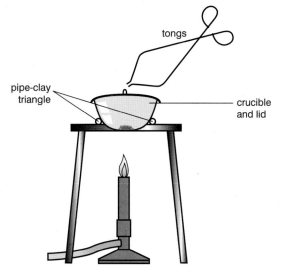

△ Fig. 6.3 Apparatus.

Mass of crucible + lid (1)	= 23.00 g
Mass of crucible + lid + magnesium (2)	= 24.13 g
Mass of crucible + lid + magnesium oxide (3)	= 24.20 g
Mass of magnesium (2 − 1)	= 0.13 g
Mass of oxygen (3 − 2)	= 0.07 g

△ Table 6.3 Results of experiment.

COMMENT

In this table of results:

✓ The description of each measurement is clear.

✓ The units are given in each case.

EXAMPLE 6

In this example a student has recorded his observations on adding various metals to dilute hydrochloric acid in a table.

Metal	Observations on adding dilute hydrochloric acid
Copper	No reaction.
Iron	Slow effervescence and a colourless gas is produced, a pale green solution forms.
Lead	A few bubbles of gas form on the surface of the metal.
Magnesium	Rapid effervescence and a colourless gas is produced, a colourless solution forms. The magnesium disappears.
Zinc	Quite rapid effervescence and a colourless gas is produced. A colourless solution is formed.

△ Table 6.4 Presenting results in a table.

COMMENT

Terms such as effervescence and solution have been used correctly and chemical names have not been included. For example, although the colourless gas referred to is hydrogen you would not actually observe hydrogen – what you actually see is effervescence. When interpreting or explaining the observations you may identify this gas as hydrogen.

4. Analysing the data and drawing conclusions

Learning objectives: to analyse and interpret data to draw conclusions from experimental activities which are consistent with the evidence, using chemical knowledge and understanding, and to communicate these findings using appropriate specialist vocabulary, relevant calculations and graphs.

Questions to ask:

What is the best way to show the pattern in my results? Should I use a bar chart, line graph or scatter graph?

✓ Graphs are usually the best way of demonstrating trends in data.

✓ A bar chart or bar graph is used when one of the variables is a **categoric variable**, for example when the melting points of the oxides of the group 2 elements are shown for each oxide the names are categoric and not continuous variables.

✓ A line graph is used when both variables are continuous, e.g. time and temperature, time and volume.

✓ Scattergraphs can be used to show the intensity of a relationship, or degree of **correlation**, between two variables.

✓ Sometimes a line of best fit is added to a scatter graph, but usually the points are left without a line.

When drawing bar charts or line graphs:

✓ Choose scales that take up most of the graph paper.

✓ Make sure the axes are linear and allow points to be plotted accurately. Each square on an axis should represent the same quantity. For example, one big square = 5 or 10 units; not 3 units.

✓ Label the axes with the variables (ideally with the independent variable on the x-axis).

✓ Make sure the axes have units.

✓ If more than one set of data is plotted use a key to distinguish the different data sets.

If I use a line graph should I join the points with a straight line or a smooth curve?

✓ When you draw a line, do not just join the dots!

✓ Remember there may be some points that don't fall on the curve – these may be incorrect or anomalous results.

✓ A graph will often make it obvious which results are anomalous and so it would not be necessary to repeat the experiment (see Example 7).

Do I have to calculate anything from my results?

✓ It will be usual to calculate means from the data.

✓ Sometimes it is helpful to make other calculations, before plotting a graph, for example you might calculate 1/time for a rate of reaction experiment.

✓ Sometimes you will have to make some calculations before you can draw any conclusions.

Can I draw a conclusion from my analysis of the results, and what chemical knowledge and understanding can be used to explain the conclusion?

✓ You need to use your chemical knowledge and understanding to explain your conclusion.

✓ It is important to be able to add some explanation which refers to relevant scientific ideas in order to justify your conclusion.

EXAMPLE 7

A student has carried out an experiment to find out how the rate of a reaction changes during the reaction. She added some hydrochloric acid to marble chips and measured the volume of carbon dioxide produced in a gas syringe. She took a reading of the volume of gas in the syringe every 10 seconds for 1.5 minutes.

The apparatus she used and the results obtained are shown below:

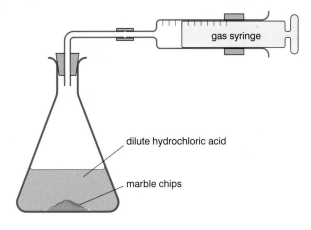

△ Fig. 6.4 Apparatus used.

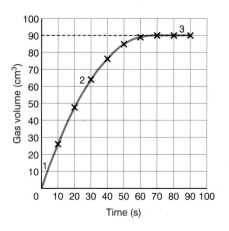

△ Fig. 6.5 Graph of experimental results.

What is the best way to show the pattern in my results?

In this experiment both the volume of gas and time are **continuous variables** and so a line graph is needed.

Straight line or a smooth curve?

With the results obtained in this experiment it is clear that a smooth curve is needed. Drawing a straight line of best fit would not be sensible.

Do I have to calculate anything from my results?

In this experiment the student had to find out how the rate of the reaction changed as the reaction proceeded. She could do this by looking at the change in steepness/gradient of the curve as the reaction proceeded and so she didn't need to do any separate calculations.

Can I draw a conclusion from my analysis of the results?
The student wrote:

As the line is steeper at the beginning of the experiment than nearer the end the rate of the reaction decreases as the reaction proceeds.

COMMENT

This is a very clear statement. In addition, the student might have referred to 'the gradient of the line' at points 1, 2 and 3 to make her conclusion even more precise.

What chemical knowledge and understanding can be used to explain the conclusion? In this investigation the student wrote:

> *As the reaction proceeds the reacting particles are converted into products. This means that there will be fewer reacting particles as time goes on, there will be fewer effective collisions and so the rate of the reaction will decrease.*

COMMENT

This is a good conclusion as it makes direct links to scientific knowledge in relation to collision theory. Reference to 'effective collisions' also indicates a good level of precision. The student might also have mentioned that as the particles react the concentration of the reactants decrease and with it the rate of the reaction.

5. Evaluating the data and methods used

Learning objective: to evaluate data and methods.

Questions to ask:

Do any of my results stand out as being inaccurate or anomalous?

✓ You need to look for any anomalous results or outliers that do not fit the pattern.

✓ You can often pick this out from a results table (or a graph if all the data points have been plotted, as well as the mean, to show the range of data).

What reasons can I give for any inaccurate results?

✓ When answering questions like this it is important to be specific. Answers such as 'experimental error' will not score any marks.

✓ It is often possible to look at the practical technique and suggest explanations for anomalous results.

✓ When you carry out the experiment you will have a better idea of which possible sources of error are more likely.

✓ Try to give a specific source of error and avoid statements such as 'the measurements must have been wrong'.

Your conclusion will be based on your findings, but must take into consideration any uncertainty in these introduced by any possible sources of error. You should discuss where these have come from in your evaluation.

Error is a difference between a measurement you make, and its true value.

The two types of errors are:

✓ random error

✓ systematic error.

With **random error**, measurements vary in an unpredictable way. This can occur when the instrument you're using to measure lacks sufficient precision to indicate differences in readings. It can also occur when it's difficult to make a measurement.

With **systematic error**, readings vary in a controlled way. They're either consistently too high or too low. One reason could be down to the way you are making a reading, e.g., taking a burette reading at the wrong point on the meniscus, or not being directly in front of an instrument when reading from it.

What an investigator *should not* discuss in an evaluation are problems introduced by using faulty equipment, or by using the equipment inappropriately. These errors can, or could have been, eliminated, by:

✓ checking equipment

✓ practising techniques before the investigation, and taking care and patience when carrying out the practical.

Overall was the method or technique I used precise enough?

✓ If your results were good enough to provide a confident answer to the problem you were investigating the method probably was good enough.

✓ If you realise your results are not precise when you compare your conclusion with the actual answer it may be you have a **systematic error** (an error that has been made in obtaining all the results.) A systematic error would indicate an overall problem with the experimental method.

✓ If your results do not show a convincing pattern then it is fair to assume that your method or technique was not precise enough and there may have been a **random error** (that is, measurements vary in an unpredictable way).

If I were to do the investigation again what would I change or improve upon?

✓ Having identified possible errors it is important to say how these could be overcome. Again you should try to be absolutely precise.

✓ When suggesting improvements, do not just say 'do it more accurately next time' or 'measure the volumes more accurately next time'.

✓ For example, if you were measuring small volumes, you could improve the method by using a burette to measure the volumes rather than a measuring cylinder.

EXAMPLE 8

A student was measuring the height of a precipitate produced in a test tube when different volumes of lead(II) nitrate solution were added to separate 15 cm^3 samples of potassium iodide solution.

$$Pb(NO_3)_2(aq) + 2KI(aq) \rightarrow PbI_2(s) + 2KNO_3(aq)$$

Do any of my results stand out as being inaccurate or anomalous?

The student plotted his results on a graph. The inaccurate result stands out from the rest. Given the pattern obtained with the other results there is no real need to repeat the result – you could be very confident that the height should have been 3.0 cm. A result like this is referred to as an anomalous result. It was an error but not a systematic error.

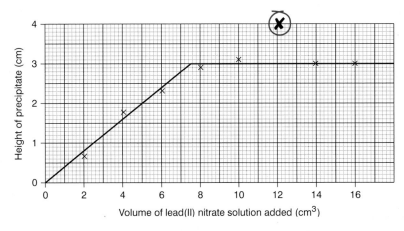

△ Fig. 6.6 Student's graph of experimental results.

What reasons can I give for any inaccurate results?

There are two main possible sources of error: either one of the volumes was measured incorrectly or the height of the precipitate was measured incorrectly.

Perhaps there was some air trapped in the precipitate and it didn't settle like the precipitates in the other tubes.

Was the method or technique I used precise enough?

You can be reasonably confident that 7.5 cm³ of lead(II) nitrate solution reacted exactly with 15 cm³ of potassium iodide solution (the point at which the height of precipitate reached its maximum value of 3.0 cm).

How can I improve the investigation?

For example, you could say, 'stir each precipitate to get rid of air bubbles and then let the precipitate settle'.

Periodic table of elements

Group 1	Group 2												Group 3	Group 4	Group 5	Group 6	Group 7	Group 0
					1 H 1 hydrogen													4 He 2 helium
7 Li 3 lithium	9 Be 4 beryllium												11 B 5 boron	12 C 6 carbon	14 N 7 nitrogen	16 O 8 oxygen	19 F 9 fluorine	20 Ne 10 neon
23 Na 11 sodium	24 Mg 12 magnesium												27 Al 13 aluminium	28 Si 14 silicon	31 P 15 phosphorus	32 S 16 sulfur	35.5 Cl 17 chlorine	40 Ar 18 argon
39 K 19 potassium	40 Ca 20 calcium	45 Sc 21 scandium	48 Ti 22 titanium	51 V 23 vanadium	52 Cr 24 chromium	55 Mn 25 manganese	56 Fe 26 iron	59 Co 27 cobalt	59 Ni 28 nickel	63.5 Cu 29 copper	65 Zn 30 zinc		70 Ga 31 gallium	73 Ge 32 germanium	75 As 33 arsenic	79 Se 34 selenium	80 Br 35 bromine	84 Kr 36 krypton
85 Rb 37 rubidium	88 Sr 38 strontium	89 Y 39 yttrium	91 Zr 40 zirconium	93 Nb 41 niobium	96 Mo 42 molybdenum	98 Tc 43 technetium	101 Ru 44 ruthenium	103 Rh 45 rhodium	106 Pd 46 palladium	108 Ag 47 silver	112 Cd 48 cadmium		115 In 49 indium	119 Sn 50 tin	122 Sb 51 antimony	128 Te 52 tellurium	127 I 53 iodine	131 Xe 54 xenon
133 Cs 55 caesium	137 Ba 56 barium	139 La 57 lanthanum	178 Hf 72 hafnium	181 Ta 73 tantalum	184 W 74 tungsten	186 Re 75 rhenium	190 Os 76 osmium	192 Ir 77 iridium	195 Pt 78 platinum	197 Au 79 gold	201 Hg 80 mercury		204 Tl 81 thallium	207 Pb 82 lead	209 Bi 83 bismuth	210 Po 84 polonium	210 At 85 astatine	222 Rn 86 radon
223 Fr 87 francium	226 Ra 88 radium	227 Ac 89 actinium																

Mathematical skills

The table below shows the mathematical skills that you will need to use during your International GCSE course. You will also need to be able to use these skills during your examinations.

			C
1	**Arithmetic and numerical computation**		
A	Recognise and use numbers in decimal form		✓
B	Recognise and use numbers in standard form		✓
C	Use ratios, fractions, percentages, powers and roots		✓
2	**Handling data**		
A	Use an appropriate number of significant figures		✓
B	Understand and find the arithmetic mean (average)		✓
C	Construct and interpret bar charts		✓
D	Understand simple probability		✓
E	Use a scatter diagram to identify a pattern or trend between two variables		✓
F	Make order of magnitude calculations		✓
3	**Algebra**		
A	Understand and use the symbols $<, >, \propto, \sim$		✓
B	Change the subject of an equation		✓
C	Substitute numerical values into algebraic equations using appropriate units for physical quantities		✓
D	Solve simple algebraic equations		✓
4	**Graphs**		
A	Translate information between graphical and numerical form		✓
B	Understand that $y = mx + c$ represents a linear relationship		✓
C	Plot two variables (discrete and continuous) from experimental or other data		✓
D	Determine the slope and intercept of a linear graph		✓
E	Understand, draw and use the slope of a tangent to a curve as a measure of rate of change		✓

Glossary

acid A substance that contains replaceable hydrogen atoms which form H^+ ions when the acid is dissolved in water. It has a pH less than 7.

acid rain Rain water that contains dissolved acids, typically sulfuric acid and nitric acid.

acidic oxide The oxide of a non-metal.

activation energy The minimum energy that must be provided before a reaction can take place.

addition polymer A polymer that is made when molecules of a single monomer join together in large numbers.

addition reaction The reaction of an alkene and another element or compound to form a single compound.

alcohol A molecule with an –OH group attached to a chain of carbon atoms.

alkali A base that is soluble in water and produces OH^- ions. It has a pH greater than 7.

alkali metal A Group 1 element.

alkane A hydrocarbon where the carbon atoms are bonded together by single bonds only.

alkene A hydrocarbon that contains a carbon-carbon double bond.

allotrope The different physical forms in which a pure element can exist.

alloy A mixture of a metal and one or more other elements.

anhydrous Literally means 'without water' – a compound, usually a salt, with no water of crystallisation.

anion A negatively charged ion, which moves to the anode during electrolysis.

anode A positively charged electrode in electrolysis.

atom The smallest particle of an element. Atoms are made of protons, electrons and neutrons.

atomic number Number of protons in an atom.

Avogadro's constant (or number) The number of particles in one mole of a substance. It is 6.0×10^{23}.

barrier method A way of preventing rusting or corrosion by covering the iron with a 'barrier' such as oil, grease or a metal such as zinc (galvanizing).

base A substance that neutralises an acid to produce a salt and water.

basic oxide The oxide of a metal.

basicity The number of replaceable hydrogen ions in a molecule of an acid.

biopolyester A polyester that can be broken down by bacteria in the soil – it is biodegradable.

boiling point The temperature of a boiling liquid – the highest temperature that the liquid can reach and the lowest temperature that the gas can reach.

bond energy The average energy associated with breaking or forming a particular covalent bond (measured in kJ/mole).

burning The reaction of a substance with oxygen in a flame.

calorimetry A method for determining energy changes in reactions or when substances are mixed together.

carbonate A salt formed by the reaction of carbon dioxide with alkalis in solution.

carbonic acid An acid formed by the reaction of carbon dioxide with water.

carboxylic acid An organic acid which contains the –COOH functional group.

catalyst A chemical that is added to speed up a reaction, but remains unchanged at the end.

catalytic cracking The process by which long-chain alkanes are broken down to form more useful short-chain alkanes and alkenes, using high temperatures and a catalyst.

cathode A negatively charged electrode in electrolysis.

cation A positive ion, which moves to the cathode during electrolysis.

cell A device for turning chemical energy into electrical energy.

chemical change A change that is not easily reversed because new substances are made.

chemical reaction A chemical change which produces new substances and which is not usually easily reversed.

chromatography The process for separating dissolved solids using a solvent and filter paper (in the school laboratory).

collision theory A theory used to explain differences in the rates of reactions as a result of the frequency and energy associated with the collisions between the reacting particles.

combustion The burning of a fuel in oxygen, also in air.

compound A pure substance formed when elements react together.

concentration Amount of chemical dissolved in 1 dm^3 of solvent.

condensation The change of state from gas to liquid.

condensation polymer A polymer formed when two monomers react together and eliminate a small molecule such as water or hydrogen chloride.

conductor A material that will allow heat or electrical energy to pass through it.

Contact process The process in which sulfuric acid is manufactured from sulfur dioxide and oxygen.

covalent bond A bond that forms when electrons are shared between the atoms of two non-metals.

cracking Forming shorter alkanes and alkenes from longer alkanes using high temperatures and a catalyst.

decomposition Chemical change that breaks down one substance into two or more.

dehydration reaction A reaction involving the removal of the elements of water from a compound.

delocalised Electrons that are not attached to a particular atom (as in a metallic structure).

diatomic Two atoms combined together (for example in a molecule).

dibromoalkane A molecule formed when bromine reacts with an alkene.

diffusion The random mixing and moving of particles in liquids and gases.

diol An alcohol with two –OH functional groups.

directly proportional The relationship between two quantities if one doubles when the other doubles.

displacement reaction A reaction in which one element takes the place of another in a compound, removing (displacing) it from the compound.

displayed formula This shows the position of all bonds and atoms in a molecule.

dissociation The splitting of a molecule to form smaller molecules or, in the presence of water, to form ions.

distillation The process for separating a liquid from a solid (usually when the solid is dissolved in the liquid).

ductile Describing a substance (such as metal) that can be drawn or pulled into a wire.

dynamic equilibrium A situation in which reactants are constantly being converted into products, and products are constantly being converted back into reactants. The rates of the forward and backward reactions are the same.

effective collision A collision with enough energy to cause a chemical reaction.

electrode The carbon or metal material that is given an electrical charge in electrolysis reactions.

electrolysis The breaking down of a compound by passing an electric current through it.

electrolyte A substance that allows electricity to pass through it when it is molten or dissolved in water.

electron Negatively charged particle with a negligible mass that forms the outer portion of all atoms.

electronic configuration The arrangement of electrons in an atom.

element A substance that cannot be broken down into other substances by any chemical change.

empirical formula The simplest formula of a compound, showing the whole number ratio of the atoms in the compound.

endothermic Type of reaction in which energy is taken in from the surroundings.

enzyme A chemical that speeds up certain reactions in biological systems, such as digestive enzymes that speed up the chemical digestion of food.

equilibrium reaction A chemical reaction where the forward and backward reactions are both likely, shown as: $X \rightleftharpoons Y$.

ester A compound formed by the reaction between an alcohol and a carboxylic acid in the presence of an acid catalyst.

evaporation When liquid changes to gas at a temperature lower than the boiling point.

exothermic A type of reaction in which energy is transferred out to the surroundings.

fermentation The process by which ethanol is made from a solution of sugar and yeast.

filtrate The clear liquid or solution produced by filtering a mixture.

formula mass (M_r) The sum of the atomic masses of the atoms in a formula.

fossil fuel Fuel made from the remains of decayed animal and plant matter compressed over millions of years.

fraction A collection of hydrocarbons that have similar molecular masses and boil at similar temperatures.

fractional distillation A process for separating liquids with different boiling points.

freezing Changing a liquid to a solid at the melting point.

fullerene A form of carbon, the most common being Buckminsterfullerene, C_{60}.

functional group A part of an organic molecule which is responsible for the characteristic reactions of the molecule.

galvanising The process of coating a metal (usually iron) with zinc.

gas The state of matter in which the substance has no fixed volume or shape.

general formula A formula, usually containing 'n' as a number (e.g. C_nH_{2n}), which enables the formula of the different members of a homologous series to be worked out.

giant covalent structure The structure of a covalently bonded molecule where multiple covalent bonds result in the formation of an extended 3D structure (e.g. in diamond).

giant ionic structure An extended 3D array of positive and negative ions producing a very strong structure with high melting and boiling points.

greenhouse gas A gas which is thought by many scientists to contribute to global warming, known as the greenhouse effect.

group A vertical column of elements in the Periodic Table.

half-equation An equation that shows the loss or gain of electrons by atoms or ions.

halogens The Group 7 elements (F, Cl, Br, I, At).

homologous series A group of organic compounds with the same general formula, similar chemical properties and physical properties that change gradually from one member of the series to the next.

hydrated Literally means 'containing water' – hydrated salts contain water of crystallisation.

hydrocarbon A compound containing only hydrogen and carbon.

indicator A substance which changes colour in either an acid or alkali, or both, and so can be used to identify acids or alkalis.

intramolecular bond A bond within a molecule.

ion A charged atom or molecule.

ionic bond A bond that involves the transfer of electrons to produce electrically charged ions.

ionic compound A compound formed by the reaction between a metal and one or more non-metals.

isomer A compound that has the same molecular formula but a different structure from a similar compound.

isotope Atoms of the same element that contain different numbers of neutrons. Isotopes have the same atomic number but different mass numbers.

kinetic theory The theory describing the movement of particles in solids, liquid and gases.

line of best fit Line on a graph that most closely matches all the data points to show a trend or pattern.

liquid The state of matter in which a substance has a fixed volume but no definite shape.

malleability The measure of how easily a substance can be beaten into sheets.

mass number The total number of protons and neutrons in an atom.

mean The arithmetic mean is the average value of a set of values. For example, the mean of the set of values 6, 5, 7 and 2 is $(6 + 5 + 7 + 2) \div 4 = 5$.

melting Changing a solid into a liquid at the melting point.

metalloid An element that has properties characteristic of both metals and a non-metals.

mixture Two or more substances combined without a chemical reaction. They may be separated easily.

molar enthalpy change The change in heat energy when the molar quantities shown in a chemical equation react together.

mole The amount of a substance containing 6×10^{23} particles (atoms, molecules, ions).

molecular formula The formula of a compound showing the actual number of atoms in it.

molecule A group of two or more atoms covalently bonded together.

monomer A small molecule that can be joined in a chain to make a polymer.

neutralisation A reaction in which an acid reacts with a base or alkali to form a salt and water.

neutron Particle present in the nucleus of atoms that have mass but no charge.

noble gas Group O elements (He, Ne, Ar, Kr, Xe, Rn). They have full outer electron shells.

non-renewable Fuel that cannot be made again in a short time span.

nucleus, atomic The tiny centre of an atom, typically made up of protons and neutrons.

order of magnitude An approximate comparison of two or more quantities in terms of powers of ten. For example, 1298 is about ten times larger than 130 (one order of magnitude larger).

ore A mineral from which a metal may be extracted.

organic chemistry The study of covalent compounds of carbon.

organic molecules Carbon based molecules.

oxidation The addition of oxygen in a chemical reaction. Electrons are lost.

oxide A product of the reaction of oxygen with another element. For example, oxygen reacts with copper to produce copper(II) oxide.

percentage yield Fraction of substance produced (actual yield) in a chemical process compared to the possible predicted yield.

period A row in the Periodic Table, from the alkali metals to the noble gases.

periodicity The gradual change in properties of the elements across each row (period) of the Periodic Table.

pH scale A scale measuring the acidity (lower than 7) or alkalinity of a solution (greater than 7). It is a measure of the concentration of hydrogen ions in a solution.

physical change A change in chemicals that is easily reversed and does not involve the making of new chemical bonds.

polymer A large molecule made up from smaller molecules (monomers). Polyethene is a polymer made from ethene.

polymerisation Making polymers from monomers.

precipitation A reaction in which an insoluble salt is formed by mixing two solutions.

products The chemicals that are produced in a reaction.

proton Positively charged, massive particles found in the nucleus of an atom.

proton acceptor A substance (base or alkali) that accepts H^+ ions (from an acid).

proton donor A substance (acid) that donates H^+ ions (to a base or alkali).

pure substance A single substance which is not mixed with any other element or compound (the opposite of a mixture).

radical An ion containing more than one atom, e.g. SO_4^{2-}.

rate of reaction How fast a reaction goes in a given interval of time.

reactant The chemicals taking part in a chemical reaction. They change into the products.

reactivity series A list of elements showing their relative reactivity. More reactive elements will displace less reactive ones from their compounds.

redox reaction A reaction involving both oxidation and reduction.

reduction When a chemical loses oxygen and gains electrons.

relative atomic mass (A_r) A number comparing the mass of one mole of atoms of a particular element with the mass of one mole of atoms of other elements. C has the value 12.

relative formula mass (M_r) The sum of the relative atomic masses of each of the atoms in one formula unit of a substance.

reversible A reaction in which reactants form products and products can form reactants.

sacrificial protection A way of preventing rusting by coating, alloying or simply putting iron into contact with a more reactive metal, such as zinc, which then reacts with oxygen in preference to the reaction of iron with oxygen.

salt Compound made from the reaction of an acid and a base or alkali.

saturated Describes an organic compound that contains only single bonds (C—C).

saturated solution A solution which cannot dissolve any more solute (solid) at that temperature.

shell A grouping of electrons around a nucleus. The first shell in an atom can hold up to 2 electrons, the next two shells can hold up to 8 each.

significant figures Digits within a measured or calculated quantity that have meaning, e.g. a measurement made with a 30 cm long ruler with divisions marked in millimetres can only have three significant figures, such as 17.4 cm; it is meaningless to state 17.42 cm because the ruler is not that precise.

simple molecular structure The covalently bonded structure of a simple molecule (as opposed to a giant molecular or covalent structure).

solid The state of matter in which a substance has a fixed volume and a definite shape.

solubility A measure of the amount of solute (solid) that dissolves in a given solution at a particular temperature, measured in g/dm^3.

solute A substance that dissolves in a solvent, producing a solution.

solution This is formed when a substance dissolves into a liquid. Aqueous solutions are formed when the solvent used is water.

solvent The liquid in which solutes are dissolved.

standard form When a number can be written in terms of powers of 10. For example, the speed of light is 300 000 000 m/s. It can also be written as 3×10^8 m/s. Standard form is useful for particularly large or small numbers.

state symbols These denote whether a substance is a solid (s), liquid (l), gas (g) or is dissolved in aqueous solution (aq).

structural formula A formula showing the main clusters of atoms in a molecule, usually including the functional group.

structural isomer A compound having the same molecular formula but different structural or displayed formula to another compound.

substitution reaction A reaction in which one atom is replaced by an atom of another element.

surface area The total area of the outside of an object. Particles of chemical reactants can bombard only the surface of an object.

tangent to a curve A tangent, in geometry, is a straight line that touches the curve at a single point and has the same direction at that point as the curve.

thermal decomposition The breaking down of a compound by heat.

titration An accurate method for calculating the concentration of an acid or alkali solution in a neutralisation reaction.

transition metal Elements found between Group 2 and 3 in the Periodic Table. Often used as catalysts and often make compounds that have coloured solutions.

universal indicator Indicating solution that turns a specific colour at each pH value.

unsaturated Describes carbon compounds that contain double bonds.

vapour Another term for gas.

volatile Easily turning to a gas.

water of crystallisation Water that occurs in crystals.

yield Amount of substance produced from a chemical reaction.

Answers

SECTION 1 PRINCIPLES OF CHEMISTRY

States of matter

Page 11

1. (I).
2. Only the solid state has a fixed shape.
3. Fine sand will pour or flow like a liquid; it takes the shape of the container it is poured into (although under a microscope you would see gaps at the surface of the container).

Page 13

1. The particles in a solid vibrate.
2. Water particles are held together more strongly in solid water.
3. Evaporation is the process in which faster-moving particles escape from the surface of a liquid.
4. A solid changes to a liquid at its melting point.

Page 18

1. Diffusion is the random mixing and moving of particles in liquids and gases.
2. The particles in the potassium manganate(VII) dissolve in the water and diffuse throughout the solution.
3. The particles of the perfume vapour/gas diffuse in the air and spread throughout the room.

Elements, compounds and mixtures

Page 26

1. Less liquid will be lost as vapour/the condenser is much more efficient at condensing the vapour as it is constantly kept cool.
2. b) – boiling points.
3. The blue dye is the most soluble (it has moved the furthest up the filter paper).
4. The retention factor $= \dfrac{\text{distance moved by the dye}}{\text{distance moved by the solvent}}$

   $= \dfrac{1.7}{10} = 0.17$
5. The boiling point will be higher than 100 °C at normal pressure.

Atomic structure

Page 32

1. The electron has the smallest relative mass.
2. Atoms have no overall charge, so the number of positive charges (protons) and negative charges (electrons) must cancel out.

Page 35

1. Isotopes are atoms of the same element with different atomic masses (different numbers of neutrons).
2. $A_r = \dfrac{(51 \times 79) + (49 \times 81)}{100} = 79.98$

The Periodic Table

Page 40

1. a) 20
   b) The atomic number is the number of protons (which equals the number of electrons) in an atom of the element. Calcium has 20 protons and 20 electrons.
   c) Group 2
   d) Period 4
   e) Calcium is a metal
2. Halogens
3. Halogens are non-metals.

Page 42

1. A metalloid has some properties typical of a metal and some of a non-metal.
2. The metal can be drawn into a wire.
3. The metal can be hammered into shape.
4. Carbon in the form of graphite will conduct electricity.
5. Non-metals usually form acidic oxides.

Page 44

1. a) 2 electrons.   b) Magnesium is in Group 2.
2. a) Aluminium

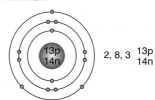

   b) Calcium

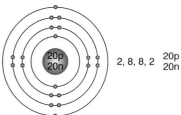

3. The noble gases have a full outer shell of electrons or have 8 electrons in their outer shell and so do not easily lose or gain electrons.

Page 45

1. Aluminium has 3 electrons in the outer shell.
2. Fluorine (F)
3. Barium (Ba)

Chemical formulae, equations and calculations

Page 52

1. a) $2Ca(s) + O_2(g) \rightarrow 2CaO(s)$
    b) $2H_2S(g) + 3O_2(g) \rightarrow 2SO_2(g) + 2H_2O(l)$
    c) $2Pb(NO_3)_2(s) \rightarrow 2PbO(s) + 4NO_2(g) + O_2(g)$
2. a) $S(s) + O_2(g) \rightarrow SO_2(g)$
    b) $2Mg(s) + O_2(g) \rightarrow 2MgO(s)$
    c) $CuO(s) + H_2(g) \rightarrow Cu(s) + H_2O(l)$

Page 53

1. a) $2C_5H_{10}(g) + 15O_2(g) \rightarrow 10CO_2(g) + 10H_2O(l)$
    b) $Fe_2O_3(s) + 3CO(g) \rightarrow 2Fe(s) + 3CO_2(g)$
    c) $2KMnO_4(s) + 16HCl(aq) \rightarrow$
    $2KCl(aq) + 2MnCl_2(aq) + 8H_2O(l) + 5Cl_2(g)$

Page 54

1. 16
2. 46
3. 48

Page 61

1. a) To allow oxygen from the air into the crucible.
    b) To prevent the loss of the magnesium oxide.
2. a) Platinum
    b) Oxygen

Page 62

1. Fe_2O_3 (0.10 mole of Fe reacts with 0.15 mole of O)
2. ZnO (0.20 mole of Zn combine with 0.20 mole of O)
3. C_4H_{10} (empirical formula mass is 29 – half of the relative formula mass)
4. H_2O_2 (empirical formula mass is 17 – half of the relative formula mass)

Page 65 (top)

1. 1 mole of $CaCO_3$ (40 + 12 + 16 + 16 +16 = 100 g) produces 1 mole of CaO (40 + 16 = 56 g)

   So 50/100 = 0.5 mole will produce 0.5 mole of CaO = 28 g
2. a) 2 moles
    b) 0.01 mole (100/1000 × 0.1)
    c) 0.25 mole (500/1000 × 0.5)

Page 65 (bottom)

1. a) 2 moles
    b) 0.5 mole
    c) 0.1 mole
2. a) 0.5 mole
    b) 0.1 mole
    c) 2 moles.

Page 66

1. 1 mole of magnesium (24 g) produces 1 mole of hydrogen gas (24 000 cm³). So 1/6 mole of Mg produces 1/6 mole hydrogen = 4 dm³ at room temperature and pressure.
2. 1 mole of oxygen gas is produced from 2 moles of hydrogen peroxide. So 0.5 mole of oxygen gas is produced from 1 mole of hydrogen peroxide.

Ionic bonding

Page 75

1.

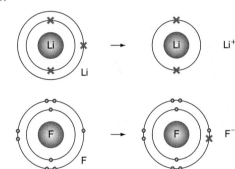

2.

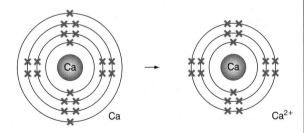

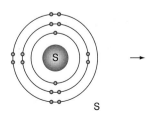

3. No. Both phosphorus and oxygen are non-metals. (A metal is needed to form an ionic bond.)

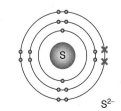

1. The ions are held together strongly in a giant lattice structure. The ions can vibrate but cannot move around.

2. Sodium chloride is made up of singly charged ions, Na$^+$ and Cl$^-$, whereas the magnesium ion in magnesium oxide has a double charge, Mg^{2+}. The higher the charge on the positive ion, the stronger the attractive forces between the positive ion and the negative ion.

Covalent bonding

Page 83

1.

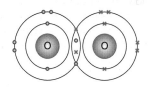

Cl – Cl

2.

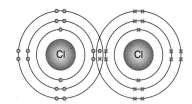

$$H-\underset{|}{\overset{H}{N}}-H$$

3.

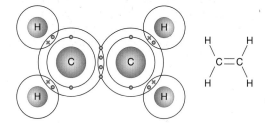

O=O

4.

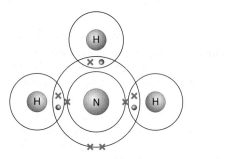

$$\underset{H}{\overset{H}{C}}=\underset{H}{\overset{H}{C}}$$

5.

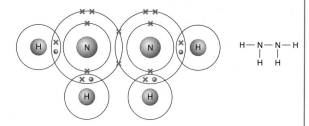

$$H-\underset{H}{\overset{}{N}}-\underset{H}{\overset{}{N}}-H$$

Page 84

1. The intermolecular forces of attraction between the molecules are weak.

2. No. There are no ions or delocalised electrons present.

Page 86

1. Each carbon atom is strongly covalently bonded to four other atoms forming a very strong giant lattice structure. A very high temperature is required to break down the structure.

2. In graphite each carbon atom is strongly covalently bonded to 3 other carbon atoms. The remaining outer shell carbon electron is delocalised and so can move along the layers formed by the covalently bonded carbon atoms.

Metallic bonding

Page 90

1. Metals contain delocalised electrons that are not fixed to a particular atom, they can move throughout the structure.

2. When a piece of metal is bent slightly the atoms slide over each other but the electrostatic attraction between the metal ions and the sea of electrons is not significantly changed.

Electrolysis

Page 96

1. The breaking down (decomposition) of a chemical compound by the use of electricity.

2. The positive electrode is the anode.

3. The substance must contain ions and they must be free to move (in molten/liquid state or dissolved in water).

Page 100

1. a) An inert electrode is an unreactive electrode. It will not be changed during electrolysis.

   b) Carbon is commonly used as an inert electrode.

2. a) Lead and chlorine

   b) Magnesium and oxygen

   c) Aluminium and oxygen

3. a) Hydrogen (sodium is above hydrogen in the reactivity series)

b) Hydrogen (zinc is above hydrogen in the reactivity series)

c) Silver (silver is below hydrogen in the reactivity series)

4. a) $2O^{2-}(aq) \rightarrow O_2(g) + 4e^-$

b) The change takes place at the anode.

SECTION 2 INORGANIC CHEMISTRY

Group 1 (alkali metals)

Page 112

1. They react with water to form alkaline solutions.

2. One electron in the outer shell.

3. The potassium atom is larger than the lithium atom so the outer electron is further from the attraction of the nucleus. It is also shielded from the nucleus by the inner shells of electrons. This outweighs the increased positive charge of the nucleus.

4. They are soft to cut (also have very low melting points).

Page 113

1. Sodium oxide is white.

2. Hydrogen. The solution formed is potassium hydroxide.

3. Sodium has one more shell of electrons than lithium and so the outer electron in sodium is further away from the attractive force of the nucleus than is the case for the outer electron in lithium. Also the extra electron shell in sodium shields the outer electron from the attractive force of the nucleus and so the electron can be more readily removed than is the case for the outer electron in lithium.

Group 7 (halogens)

Page 120

1. The melting point increases from chlorine to iodine.

2. Astatine would be a solid at room temperature. The trend down the group is from gas (fluorine, chlorine) to liquid (bromine) to solid (iodine). Hence astatine would be expected to be a solid at room temperature.

3. Seven electrons in the outer shell.

4. The atoms only need to gain one electron to achieve 8 in the outer shell.

5. The chlorine molecule is made up of two atoms combined/bonded together, Cl_2.

6. A displacement reaction for the Group 7 elements is one in which a more reactive halogen displaces a less reactive halogen from a solution of one of its salts. For example, chlorine will displace bromine from a solution of sodium bromide.

7. The displacement reaction involves one Group 7 element being reduced (gaining electrons) and one being oxidised (losing electrons).

Page 122

1. a)

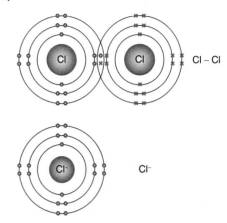

b) $2Cl^-(aq) \rightarrow Cl_2(g) + 2e^-$. The chloride ions lose electrons.

c) $2NaOH(aq) + Cl_2(aq) \rightarrow$
$$NaCl(aq) + NaClO(aq) + H_2O(l)$$

d) Chlorine is a more reactive halogen than bromine. Chlorine will displace bromine from a solution of a bromide ions. (Chlorine will oxidise bromide ions to bromine.)

Observations: chlorine water is pale green. When this is added to a colourless solution of potassium bromide the resulting solution will turn orange due to the presence of bromine.

2. a) The reactivity of fluorine is due to its electronic structure 2,7. Fluorine only needs to gain one electron to form a fluoride ion. This is very easy because of its small size and high attractive force of the nucleus.

b) $F_2 + 2e^- \rightarrow 2F^-$

c) Both chlorine and iodine are less reactive than fluorine. Fluorine could only be displaced from fluoride ions by a more reactive halogen. As there are no halogens that are more reactive than fluorine, fluorine will not be displaced from fluoride ions.

Gases in the atmosphere

Page 128

1. Argon

2. 0.04%

3. Nitrogen

4. Copper(II) oxide

1. A basic oxide will react with an acid to form a salt.

2. $2Ca(s) + O_2(g) \rightarrow 2CaO(s)$

3. Acidic oxide. Phosphorus is a non-metal.

Reactivity series

Page 137

1. No. Copper is below hydrogen in the reactivity series.

2. $2K(s) + 2H_2O(l) \rightarrow 2KOH(aq) + H_2(g)$

3. $Mg(s) + PbO(s) \rightarrow MgO(s) + Pb(s)$

4. No. Carbon is below magnesium in the reactivity series.

5. The magnesium atoms are oxidised. They lose electrons and form magnesium ions.

Page 139

1. Air (oxygen) and water must be present.

2. The grease can be easily removed or wiped away.

3. a) Galvanising involves coating iron or steel with zinc.

   b) As zinc is more reactive than iron, moist air will react with zinc in preference to the iron.

Page 139

1. a) oxygen + water

   b) chromium protects iron from oxygen and water/ used as it is shiny – good decorative effect

2. Aluminium forms aluminium oxide which acts as a protective layer – does not react with the air.

Extraction and uses of metals

Page 147

1. Gold or silver

2. Potassium, sodium, calcium or magnesium

3. Aluminium is above carbon in the reactivity series and so carbon cannot reduce aluminium oxide to aluminium.

4. a) The cryolite produces an electrolyte that has a lower melting point than that of the aluminium oxide and so reduces energy costs.

   b) Aluminium atoms form at the cathode.

   c) $Al^{3+}(l) + 3e^- \rightarrow Al(l)$

   d) The aluminium ions are reduced. They accept electrons.

   e) The carbon electrodes react with the oxygen produced in the electrolysis and form carbon dioxide.

5. Aluminium has a high strength to weight ratio/low density/resists corrosion.

1. a) $2Al(s) + Fe_2O_3(s) \rightarrow 2Fe(s) + Al_2O_3(s)$

   b) Aluminium is higher in the reactivity series than iron therefore it is more reactive. Aluminium is able to displace the less reactive iron from its oxide and so form iron and aluminium oxide.

   c) Any metal which is higher than iron in the reactivity series can be selected. The higher the metal in the series the more reactive the metal and the more reactive the reaction will be. If the chosen metal is above aluminium, the reaction is more reactive.

   d) Aluminium is displacing iron in iron(III) oxide and becoming aluminium oxide by losing electrons. Iron(III) oxide is gaining electrons to become iron metal.

   Redox is when oxidation and reduction occur. Aluminium is losing electrons (oxidation). Iron(III) oxide is gaining electrons (reduction).

Page 153 (top)

1. a) Carbon is more reactive than iron, so it can push out or displace the iron from iron oxide.

   b) $2Fe_2O_3(s) + 3C(s) \rightarrow 4Fe(s) + 3CO_2(g)$

2. a) An alloy is a mixture of a metal and another element.

   b) The proportion of carbon is reduced by heating the iron produced by the blast furnace in oxygen.

   c) The iron produced by the blast furnace is brittle – steel is more flexible and more resistant to corrosion.

Page 153 (bottom)–154

1. a) Haematite/iron ore (Fe_2O_3) and bauxite (Al_2O_3).

   b) Carbon is not a strong enough reducing agent to remove titanium from its ore. (Note: titanium does form titanium carbide TiC if it is heated with carbon.)

   c) i) titanium(IV) oxide + chlorine + carbon(coke) → titanium (IV) chloride + carbon monoxide

      $TiO_2(s) + 2Cl_2(g) + 2C(s) \rightarrow TiCl_4(l) + 2CO(g)$

      ii) titanium(IV) chloride + sodium → titanium + sodium chloride

      $TiCl_4(l) + 4Na(s) \rightarrow Ti(s) + 4NaCl(s)$

      iii) Displacement/redox

      iv) Sodium is a reactive metal and will displace titanium from its chloride. Sodium is acting as a reducing agent. Sodium metal will lose electrons to form sodium ions (sodium chloride). Titanium ions will gain electrons to become titanium.

      v) Sodium is a very reactive metal (reacts with water and air) the reaction therefore needs to be carried out in an inert atmosphere. Magnesium is still more reactive than titanium and will displace titanium from titanium chloride, but magnesium is not as reactive as sodium so absence of water/air is not as important. (Note, however, $TiCl_4$ does react violently with water.)

Reaction $TiCl_4(l) + 2Mg(s) \rightarrow Ti(s) + 2MgCl_2(s)$

vi) Copper is low in the reactivity series. It will not displace titanium from its compounds. Copper is not a strong enough reducing agent to remove titanium from its ore.

Acids, alkalis and titrations

Page 160

1. Both solutions are alkalis. Solution A is weakly alkaline whereas solution B is strongly alkaline.

2. The solution is acidic.

3. Calcium is a metal. The oxides (and hydroxides) of metals are bases.

Page 163

1. Neutralisation is the reaction between an acid and an alkali or a base to form a salt and water.

2. $H^+(aq)$

3. $OH^-(aq)$

Acids, bases and salt preparations

Page 171

1. A salt is formed when a replaceable hydrogen of an acid is replaced by a metal.

2. Sulfuric acid

3. Potassium chloride will be soluble in water (as are all potassium salts).

4. Calcium nitrate

Page 172

1. Precipitation is the formation of an insoluble salt as a result of a chemical reaction taking place in aqueous solution.

2. Filtration

3. Washing with cold water will remove traces of any remaining soluble salts.

4. **a)** lead(II) nitrate + sodium chloride → lead(II) chloride + sodium nitrate

   b) $Pb(NO_3)_2(aq) + 2NaCl(aq) \rightarrow PbCl_2(s) + 2NaNO_3(aq)$

Page 173

1. **a)** Magnesium carbonate + sulfuric acid → magnesium sulfate + carbon dioxide + water

   $MgCO_3(s) + H_2SO_4(aq) \rightarrow MgSO_4(aq) + CO_2(g) + H_2O(l)$

   Note: Magnesium sulfate also known as Epsom Salts is $MgSO_4 . 7H_2O$ (water could be added to both sides of the equation if this formula was required).

b) Any risk assessment format could be used: ensure

Hazard: Risk: Precautions are included

All chemicals to be included (both reactants and products)

All glassware

Basic laboratory safety may be included but this could be listed or the lab rules referred to.

c) Equipment: beaker ($100\ cm^3/250\ cm^3$); measuring cylinder ($250\ cm^3$); spatula; glass rod; filter paper and funnel; stand and clamp for filtering/tripod and gauze; Bunsen burner; evaporating basin

d) Instructions

- Complete and follow risk assessment during experiment.
- Measure ($50\ cm^3$) dilute sulfuric acid into a beaker.
- Using a spatula add magnesium carbonate and stir the mixture. Continue adding the magnesium carbonate until fizzing stops and the magnesium carbonate is in excess.
- Filter the excess magnesium carbonate and collect the magnesium sulfate in an evaporating basin.
- Heat the solution in the evaporating basin until about half of the water has been removed. Leave the solution at room temperature and crystals of magnesium sulfate should form.

e) The white powder is anhydrous magnesium sulfate – this is magnesium sulfate without the water of crystallisation. The presence of the water gives the crystalline structure, if this water is removed a powder is produced.

f) In order to prepare a salt using an acid–base titration method both reactants need to be solutions. Magnesium carbonate is not soluble in water and therefore a standard solution can not be prepared.

g) Anhydrous form formula $MgSO_4$. It can act as a drying agent as it readily absorbs water from the air. (It is hygroscopic.)

Chemical tests

Page 180

1. Calcium ions produce an orange red flame colour.

2. A nichrome wire is used in a flame test.

3. Add sodium hydroxide solution. Fe^{2+} produces a green precipitate; Fe^{3+} produces a reddish brown precipitate.

Page 182

1. Add dilute sodium hydroxide and heat. An alkaline gas (turns red litmus paper blue) indicates the presence of an ammonium compound.

2. **a)** Carbon dioxide

   b) Bubble the gas through lime water. A white precipitate forms.

3. The Fe^{3+} ion is present in solution X.

4. The Cl^- ion is present in solution Y.

Page 183

1. Ammonia

2. Oxygen

3. Chlorine

Page 184

1. **a)** Dip wire in concentrated hydrochloric acid, heat strongly in flame until absence of any yellow colour (sodium)

   b) Any three from:

Name of cation	Colour of precipitate
zinc/lead	white
magnesium/calcium	white
copper(II)	blue
iron(II)	dark mud green/ turns brown slowly
iron(III)	rust brown/orange

   c) $Ag^+ (aq) + X^- (aq) \rightarrow AgX(s)$ where X^- is Cl^-, Br^-, I^-.

   d) HCl is added to remove any carbonate ions that may be present.

2. Plan needs to check for testing of both anion and cation for each sample and should include practical instructions.

   Blue compound

   Test for copper(II) – sodium hydroxide : result blue precipitate

   $Cu^{2+}(aq) + 2OH^-(aq) \rightarrow Cu(OH)_2(s)$

   Test for sulfate – hydrochloric acid/barium chloride : result white precipitate

   $Ba^{2+}(aq) + SO_4^{2-}(aq) \rightarrow BaSO_4(s)$

   White compound

   Flame test for Na^+ – yellow

   Test for carbonate – add dilute acid – effervescence/carbon dioxide evolved – turns lime water milky

   $CO_3^{2-}(s) + 2H^+(aq) \rightarrow H_2O(l) + CO_2(g)$

Page 185

1. Without water/without water of crystallisation.

2. Determine the boiling point of the liquid. (Pure water boils at 100 °C at normal pressure.)

SECTION 3 PHYSICAL CHEMISTRY

Energetics

Page 200

1. A reaction that releases heat energy to the surroundings.

2. A reaction that absorbs energy from the surroundings.

3. Polystyrene is a very good insulator and so very little energy is transferred to the surroundings.

4. Energy change = $100 \times 4.2 \times 6 = 2520$ J

Page 201

1. A high proportion of the energy released is transferred to the surrounding air.

2. Energy = $30 \times 4.2 \times 15 = 1890$ J for 0.1 g corn puffs. Energy for 1 g corn puffs = $1890 \times 10 = 18\,900$ J

3. Energy = $200 \times 4.2 \times 14 = 11\,760$ J. Energy per mole = $11\,760 \times 46/0.5 = 1\,081\,920$ J (1082 kJ)

Page 202

1. Endothermic

2. The activation energy

Page 206

1. The sign indicates whether the reaction is exothermic (negative sign) or endothermic (positive sign).

2. Energy is needed to break bonds.

3. In an endothermic reaction, more energy is needed to break bonds than is recovered on forming bonds.

4. The units are kJ/mol.

5. $2H_2(g) + O_2(g) \rightarrow 2H_2O(l)$

   Energy needed to break bonds = $(2 \times 436) + 498 = +1370$ kJ

   Energy released on forming bonds = $4 \times 464 = -1856$ kJ

   Energy change = $(+1370) - (1856) = -486$ kJ

Page 208

1. **a)** $2H_2(g) + O_2(g) \rightarrow 2H_2O(l)$

   b) In an exothermic reaction energy is transferred to the surroundings and the temperature rises.

   c) Exothermic reaction

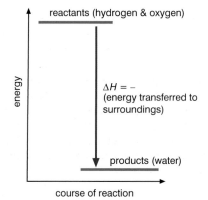

d)

Bonds broken (energy absorbed)		Bonds formed (energy released)	
O=O	498	O–H×4	464×4
H–H×2	436×2 = 872		
Total	1370		1856

Overall energy change 1856 – 1370 = 486 kJ

Equation $2H_2(g) + O_2(g) \rightarrow 2H_2O(l)$

Therefore overall energy change is 486/2 = –243 kJ/mol of hydrogen burned.

2. At present it is expensive to manufacture hydrogen by using electrolysis of water.

(Although obtaining hydrogen from electrolysis using wind or solar generated electricity is now being considered.)

Hydrogen also requires much more storage volume than the volume of petrol that produces the equivalent energy.

Hydrogen must be stored carefully as it is explosive in air.

Rates of reaction

Page 214

1. • The particles must collide.
   • There must be sufficient energy in the collision (to break bonds).

2. An effective collision is one which results in a chemical reaction between the colliding particles.

3. It is an energy barrier. Only collisions which have enough energy to overcome this barrier will lead to a reaction.

Page 216

1. A gas syringe will accurately measure the volume of gas produced.

2. No gas is being produced – the reaction hasn't started or it is finished.

3. The quicker reaction will have the steeper gradient.

Page 221

1. The units of concentration for solutions are mol/dm³.

2. The particles are more closely packed together and so there will be more (effective) collisions per second.

3. Increasing temperature means the particles: have more (kinetic) energy; more of the collisions will have energy greater than or equal to the activation energy; there will be more effective/successful collisions per second as the particles are moving faster and so will collide more often and with more energy in the collisions.

Page 225 (top)

1. A catalyst is a substance that changes the rate of a chemical reaction and remains chemically unchanged at the end of the reaction.

2. A biological catalyst is called an enzyme.

3. Changes in pressure affect reactions involving gases.

Page 225 (bottom)

1. **a)** Magnesium + hydrochloric acid → magnesium chloride + hydrogen

   $Mg(s) + 2HCl(aq) \rightarrow MgCl_2(aq) + H_2(g)$

   b) i)

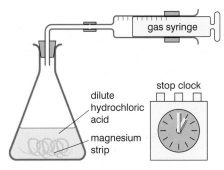

   ii) Length of magnesium strip; concentration and volume of hydrochloric acid.

   iii) In experiment 1 the curve is steeper (has a greater gradient) than in experiment 2.

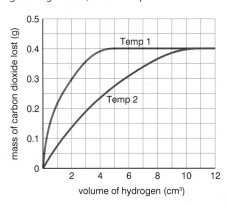

   Experiment 1 – is at the higher temperature. Hydrogen is evolved at a faster rate. Reaction complete as shown (indicated as the volume of gas stays constant).

   Experiment 2 – gradient of graph less showing hydrogen gas evolved at a slower rate. Reaction complete as shown.

2. The initial rate of each of the chemical reactions can be found by drawing a gradient to the curve at the starting point (time = 0 seconds).

   A gradient needs to be drawn at the start of the reaction – gradient found as change of y-axis/

change in *x*-axis :

Change in volume/unit time or loss in mass/unit time

The initial rate of the reactions for the experiments performed at the same temperatures should be the same as long as the same length of magnesium/ concentration and volume of hydrochloric acid are used. The initial rate of the experiment at the higher temperature will be greater as long as the conditions were the same.

Reversible reactions and equilibria

Page 232

1. The copper(II) sulfate crystals can be converted into anhydrous copper(II) sulfate by heating and removing the water in the crystals. When water is re-added to anhydrous copper(II) sulfate the copper(II) sulfate crystals are re-formed.

2. The concentration of each of the products and reactants remains constant unless a change is made to the reaction.

3. Reactants are constantly being converted into products and products are constantly being converted back into reactants. The rates of these two reactions are the same.

Page 235

1. a) Low temperature
    b) High pressure
    c) Using a catalyst has no impact on the position of equilibrium.

2. A catalyst will increase the rate of the forward and backward reactions to the same extent and so the equilibrium composition of reactants and products will not be changed.

SECTION 4 ORGANIC CHEMISTRY

Crude oil

Page 247

1. The supplies of crude oil are limited – it takes millions of years for crude oil to be formed.

2. Natural gas or methane. It is trapped in pockets above the oil.

3. Small chain of carbon atoms.

4. Long chain of carbon atoms.

5. These fractions readily form a vapour.

Page 249

1. Ethene is a member of the alkene homologous series.

2. The fractional distillation of crude oil produces a high proportion of long-chain hydrocarbons, which are not as useful as short-chain hydrocarbons. Cracking converts the long-chain hydrocarbons into more useful shorter chain hydrocarbons.

3. The conditions required for cracking oil fractions are a temperature of between 600 and 700 °C and a catalyst of silica or alumina.

Page 251

1. The fuel will burn with a yellow (rather than blue) flame.

2. Carbon and carbon monoxide.

3. It combines with the haemoglobin in the blood to form carboxyhaemoglobin, which prevents the haemoglobin from combining with oxygen. This starves the cells in the body of the oxygen they need to function.

4. Wind, wave, solar, geothermal, hydroelectric and nuclear power are alternative ways of generating energy.

Page 253

1. The sulfur dioxide combines with oxygen and water to form sulfuric acid.

2. Nitrogen oxide reacts with oxygen and water to form nitric acid.

3. An alkali such as slaked lime.

Alkanes

Page 262

1. a) Contains no C=C double bonds.
    b) A compound containing hydrogen and carbon only.

2. a) C_5H_{12}
    b) Carbon dioxide and water.

3. Hexane will be a liquid. Its physical properties will most closely resemble those of pentane.

4. There is no real functional group for the alkanes – single C–C and C–H bonds are common in all organic chemicals.

Page 264 (top)

1. Isomers are molecules with the same molecular formula but different structural formulae.

2.

$CH_3CH_2CH_2CH_2CH_2CH_3$

The molecular formula is C_6H_{14}.

3. Hexane. Any two isomers, for example (accept any correctly drawn isomers.)

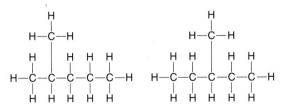

Page 264 (bottom)

1. Fuels are substances that provide heat energy. Alkanes burn readily in air combining with oxygen to produce carbon dioxide and water vapour and large quantities of heat.

2. Incomplete combustion leads to the formation of carbon monoxide instead of carbon dioxide. It is a very poisonous gas and particularly dangerous as it has no odour and causes drowsiness. Carbon monoxide is poisonous because it reacts with the haemoglobin in the blood, forming 'carboxyhaemo-globin'. The haemoglobin is no longer available to transport oxygen to the body and death results from oxygen starvation.

3. Short-chain hydrocarbons are more likely to form carbon dioxide and water as their main products as there is less carbon per molecule in these hydrocarbons to react with the available oxygen. Longer chain hydrocarbons often burn with a smoky flame and leave black carbon deposits as there is insufficient oxygen to form carbon dioxide, with the many carbons in the longer chains. Carbon monoxide is also formed.

Alkenes

Page 271

1. It contains at least one C=C double bond.

2. The manufacture of polymers (polyethene).

3. a) $CH_2=CHCH_3$

b)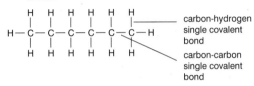

Page 272

1. a) Saturated compounds contain only covalent single bonds. Alkanes contain only carbon-carbon and carbon-hydrogen single bonds and are therefore saturated hydrocarbons. Alkenes contain C=C double bond and are therefore unsaturated.

b) Alkane: C_6H_{14}, alkene C_6H_{12}. The position of the double bond can be between any pair of carbon atoms, but there should be one less hydrogen attached to each of the double-bonded carbons.

hexane

carbon-hydrogen single covalent bond

carbon-carbon single covalent bond

hexene

carbon-hydrogen single covalent bond

carbon-carbon single covalent bond

carbon=carbon double covalent bond

c) Isomers of hexane

2-methylpentane

2,2-dimethylbutane

d) Hexane undergoes a substitution reaction with bromine (any H can be substituted with Br/can have more than one substitution). HBr is also formed in the reaction. The diagram shows 1-bromohexane.

Hexene undergoes an addition reaction with bromine (note bromine loses its colour – this is a test for unsaturation). The bromine will add across the double bond. Example shows addition of hex-1-ene to form 1,2-dibromohexane.

Alcohols

Page 277

1. C_4H_9OH

2. It is a relatively 'clean' fuel and releases only carbon dioxide and water into the atmosphere. (It does not release sulfur dioxide and nitrogen oxides, as petrol does when it burns.)

3. A solvent is a liquid that dissolves other substances (solutes) to form solutions.

Page 278

1. Fermentation is the process in which ethanol is made from sugar, yeast and water.

2. The optimum temperature is in the range 25 to 30°C.

3. The yeast contains enzymes which increase the rate of the reaction.

4. Carbon dioxide.

Page 283

1. a) $CH_2=CH_2(g) + H_2O(g) \rightleftharpoons$
        $C_2H_5OH(g)$ ($\Delta H = -45$ kJ/mol)

    Any reaction in which heat is given out by a system to the surroundings is called an exothermic reaction.

    b) The reaction for the formation of ethanol is reversible. The forward reaction is exothermic and therefore low temperatures are favoured in order to produce maximum product (i.e. to shift the position of equilibrium to the right). Low temperatures however slow down the rate of the reaction. Manufacturers are trying to produce as much ethanol as possible each day and therefore a compromise temperature is used to balance the amount of ethanol

produced with the time it takes. Although high temperatures would increase the rate, they would produce a poor yield of ethanol.

   c) In the reaction for the formation of ethanol from ethene and steam there are two molecules on the left hand side of the equation but only one on the right, therefore the formation of ethanol is favoured by high pressure. High pressures also increase the rate of reaction, but high pressures are expensive as the plant needs strong pipes and containment vessels and high energy requirements. High pressure may also cause the ethene to polymerise. Pressures of 60–70 atmospheres are therefore used.

   d) Although adding phosphoric(V) acid as a catalyst does not produce any greater percentage of ethanol in the equilibrium mixture, it does ensure that the reaction is fast enough for the dynamic equilibrium to be set up within the time the gases are within the reactor. Without the catalyst the reaction would be too slow to make the manufacture of ethanol cost-effective.

Synthetic polymers

Page 297

1. The individual beads are like monomer molecules. The string of beads is like a polymer made by joining together many of these monomers.

2. Poly(ethene)/polythene is used to make plastic bags.

3. a)

   b)

4. Poly(chloroethene) is an addition polymer.

Page 299

1. Polyester is made from two monomers and a small molecule is eliminated when the two monomers combine. An addition polymer typically has only one monomer.

2. To produce a polymer chain, the monomers need to be able to join together at both ends of the molecules.

INDEX

Acknowledgements

Every effort has been made to trace copyright holders and to obtain their permission for the use of copyright material. The publishers will gladly receive any information enabling them to rectify any error or omission at the first opportunity. The publishers would like to thank the following for permission to reproduce photographs:

(t = top, b = bottom, c = centre, l = left, r = right)

Cover & p1 Sean Lema/Shutterstock, p8-9 Denis Vrublevski/Shutterstock, p10 Jele/Shutterstock, p11 Achim Baque/Shutterstock, p15 Cobalt88/Shutterstock, p16 3x Andrew Lambert Photography/Science Photo Library, p22 Fikmik/Shutterstock, p30 Scott Rothstein/Shutterstock, p33 David Parker & Julian Baum/Science Photo Library, p44 Ho Philip/Shutterstock, p48 Martyn F. Chillmaid/Science Photo Library, p50 Leslie Garland Picture Library/Alamy, p52 Leslie Garland Picture Library/Alamy, p54 Andrew Lambert Photography/Science Photo Library, p55 Science Photo Library, p56 Arena Creative/Shutterstock, p57 Andrew Lambert Photography/Science Photo Library, p60 Andrew Lambert Photography/Science Photo Library, p72 Smit/Shutterstock, p74 Brian Weed/Shutterstock, p76 Blaz Kure/Shutterstock, p77t Travis Manley/Shutterstock, p77b Marc Dietrich/Shutterstock, p80 Voronin76/Shutterstock, p85l Broukoid/Shutterstock, p85r Tyler Boyes/Shutterstock, p86t Dmitry Kalinovsky/Shutterstock, p86b Tatiana Grozetskaya/Shutterstock, p89 Demarcomedia/Shutterstock, p91 Feraru Nicolae/Shutterstock, p94 Maximilian Stock Ltd/Science Photo Library, p108-109 Piotr Zajc/Shutterstock, p110 Andrew Lambert Photography/Science Photo Library, p111 Charles D. Winters/Science Photo Library, p113 Andrew Lambert Photography/Science Photo Library, p114 Design56/Shutterstock, p117 Andrew Lambert Photography/Science Photo Library, p126 Andrew Lambert Photography/Science Photo Library, p128 Martyn F. Chillmaid/Science Photo Library, p129 CSU Archv/Everett/Rex Features, p130 Justin S/Shutterstock, p134 Mypix/Shutterstock, p138 Joe Gough/Shutterstock, p143 Holly Kuchera/Shutterstock, p144 David.Monniaux, p147 Rtimages/Shutterstock, p148t Kilukilu/Shutterstock, p148l Fokin Oleg/Shutterstock, p148r Parnumas Na Phatthalung/Shutterstock, p149l Holly Kuchera/Shutterstock, p149r Danicek/Shutterstock, p151 Mffoto/Shutterstock, p158 Martyn F. Chillmaid/Science Photo Library, p161 Jcwait/Shutterstock, p167 Blaz Kure/Shutterstock, p168 Andrew Lambert Photography/Science Photo Library, p170 Charles D. Winters/Science Photo Library, p178 Monica Wilde www.angelflames.com, p179 Andrew Lambert Photography/Science Photo Library, p181 Andrew Lambert Photography/Science Photo Library, p185 Martyn F. Chillmaid/Science Photo Library, p196-97 Galyna Andrushko/Shutterstock, p198 Badahos/Shutterstock, p207t Iakov Kalinin/Shutterstock, p207b John T. Fowler/Alamy, p212 Anna Baburkina/Shutterstock, p222l Dgmata/Shutterstock, p222r Ken Brown/iStockphoto, p222b Martyn F. Chillmaid/Science Photo Library, p223 Charles D. Winters/Science Photo Library, p230 Martyn F. Chillmaid/Science Photo Library, p231 Andrew Lambert Photography/Science Photo Library, p234 Sciencephotos/Alamy, p242-43 Beaucroft/Shutterstock, p244 Paul Rapson/Science Photo Library, p245 Bestweb/Shutterstock, p250 Joe Gough/Shutterstock, p252 Mark Edwards/Still Pictures, p254 Dawid Zagorski/Shutterstock, p259 Yvan/Shutterstock, p260 Speedpix/Alamy, p261 Leslie Garland Picture Library/Alamy, p262 Rick Decker/Alamy, p266 Muzsy/Shutterstock, p269 Jolin/Shutterstock, p271 Africanstuff/Shutterstock, p275 Tonis Valing/Shutterstock, p277 David R. Frazier Photolibrary, Inc./Alamy, p282 Jim Parkin/Alamy, p286 Goran Bogicebic/Shutterstock, p290 Volosina/Shutterstock, p295 Dmitry Yashkin/Shutterstock, p297 Kaband/Shutterstock, p298 Epicstockmedia/Shutterstock, p299 Green Stock Media/Alamy, p300 Lya_Cattel/iStockphoto, p318 Ed Phillips/Shutterstock

NOTES

NOTES

NOTES